P9-AQN-501

Physicists in Conflict

Physicists in Conflict

Neil A Porter

Institute of Physics Publishing
Bristol and Philadelphia

British Library Cataloguing-in-Publication Data

A catalogue record for this book is available from the British Library.

ISBN 0 7503 0509 6

Library of Congress Cataloging-in-Publication Data are available

Published by Institute of Physics Publishing, wholly owned by The Institute of Physics, London

Institute of Physics Publishing, Dirac House, Temple Back, Bristol BS1 6BE, UK

US Office: Institute of Physics Publishing, Suite 1035, The Public Ledger Building, 150 South Independence Mall West, Philadelphia, PA 19106, USA

Printed in the UK by J W Arrowsmith Ltd, Bristol

For Sheila and Rahel

CONTENTS

PREFACE

This is not a book about conflict in physics, but rather about individuals caught up in conflict because of their physics. Obviously one cannot separate the two, and some remarks are made in the Introduction (Chapter 1) and Conclusions (Chapter 13) about the nature and consequences of conflict. Not everyone agrees that conflict is bad. Richard Feynman is quoted as saying in 1973: 'When everything is so neatly wrapped up ... with all experiments in exact agreement with each other and with the theory ... one is learning absolutely nothing. On the other hand, when experiments are in hopeless conflict—or when the observations do not make sense according to conventional ideas, or when none of the new models seems to work—one is really making progress and a breakthrough is just around the corner'. One might distinguish between results in conflict and authors in conflict.

Perhaps one might distinguish also between benign and malignant conflict. Benign conflict, which I prefer to call dialogue, is epitomized by the conflict between Bohr and Einstein. Here were protagonists who respected and admired each other, and their controversy, though somewhat confused, led to a better understanding of quantum mechanics. It may be significant that both sides had the same ultimate aim, the advancement of physics. In the Galileo case, however, one side was concerned with the advancement of astronomy, the other with the preservation of a literal interpretation of Scripture and the defence of Aristotelian doctrine. It was entirely malignant and led to the virtual destruction of science in Italy. There is a spectrum of conflict between these two instances, between dialogue and persecution.

What I seek to show in this book is that:

(i) scientific conflict is different in some respects from other forms of conflict;
(ii) scientific conflicts, particularly with authority, resemble each other more than they resemble other forms of conflict.

I must thank the following people for reading or commenting on part or all of the manuscript: Ann Breslin, Stephen Fegan, William

Mathews, Brian Riley and Trevor Weekes; and for their encouragement, Paul Lennon and other staff of the Milltown Institute. Remaining errors of fact or erroneous opinion are my own.

Neil Porter
May 1998

ACKNOWLEDGMENTS

The author gratefully acknowledges permission to quote excerpts from the following works:

Reid R W 1959 *Tongues of Conscience*. Copyright © 1959 Constable and Co. Reprinted with permission from the publisher, Constable and Co.

Oppenheimer J R 1954 *The Trial: In the Matter of J Robert Oppenheimer. Transcript of Hearings before the Personnel Security Board, US Atomic Energy Commission*. Reprinted with permission from the US National Archive.

Brush S G 1964 *Statistical Physics and the Atomic Theory of Matter*. Copyright © 1964 Princeton University Press. Reprinted with permission from Princeton University Press.

Redondi P 1987 *Galileo Heretic*. Copyright © 1987 Princeton University Press. Reprinted by permission from Princeton University Press.

Finocchiaro M A 1989 *The Galileo Affair*. Copyright © 1989 Regents of the University of California. Reprinted with permission from California University Press.

Pais A 1982 *Subtle is the Lord. The Science and the Life of Albert Einstein*. Copyright © 1982 Oxford University Press. Reprinted by permission from Oxford University Press.

Imerti A (ed) 1964 *The Expulsion of the Triumphant Beast* by Giordano Bruno. Published by Rutgers University Press, © 1978 A Imerti. Reprinted by permission.

Dzielska M 1995 *Hypatia of Alexandria*. Translated by F Lyra. Copyright © 1995 by the President and Fellows of Harvard College. Reprinted by permission of Harvard University Press.

Clark P 1978 Atomism and thermodynamics (in Howson (ed) *Appraisal in the Physical Sciences*). Copyright © 1978 Cambridge University Press. Reprinted with permission from Cambridge University Press.

de Santillana G 1958 *The Crime of Galileo*. Copyright © 1958 Heinemann.

Rhodes R 1986 *The Making of the Atomic Bomb* (London: Penguin).

Dyson F J 1995 *From Eros to Gaia* (London: Penguin).

Snow C P 1971 *Science and Government in Public Affairs* (London: Macmillan).

Weinberg S 1993 *Dreams of a Final Theory* (London: Vintage)

The copyright owners of five sources could not be traced.

The author gratefully acknowledges permission to use the following illustrations:

Figures 2.1, 2.2, and 3.3, by courtesy of Cambridge University Press.
Figure 4.2, by courtesy of Chatto and Windus Ltd.
Figure 7.3, by courtesy of Oxford University Press.
Figure 10.3, by courtesy of the Royal Astronomical Society.

Photographs

2.3 Roger Bacon sends his manuscript secretly to the Pope, from prison. Unnamed artist in Figuier, Les Grand Savants. By courtesy of the Mary Evans Picture Library.
3.1 Nicholas Copernicus. By courtesy of AIP Emilio Segre Visual Archives.
3.2 Ptolemy's planetary system which features the Earth at the centre of the universe. *Harmonia Macrocosmica* of Andreas Cellarius. 1650. By courtesy of the Mary Evans Picture Library.
3.4 Saint Robert Bellarmine. By courtesy of the Mary Evans Picture Library.
3.5 Galileo Galilei. By courtesy of AIP Emilio Segre Visual Archive.
4.1 Johannes Kepler. By courtesy of the Mary Evans Picture Library.
5.1 William Thomson, Lord Kelvin. By courtesy of the Mary Evans Picture Library.
5.2 James Clerk Maxwell. By courtesy of the University of Cambridge, Cavendish Laboratory, Madingley Road, Cambridge, England.
5.3 Ernst Mach. By courtesy of AIP Emilio Segre Visual Archives, Physics Today Collection.
5.4 Ludwig Boltzmann. University of Vienna. By courtesy of AIP Emilio Segre Visual Archives.
6.1 Rene Blondlot. By courtesy of AIP Emilio Segre Visual Archives, Physics Today Collection.
7.1 Albert Einstein. By courtesy of AIP Emilio Segre Visual Archives.
7.2 Niels Bohr. Princeton University. By courtesy of AIP Emilio Segre Visual Archives.
8.1 J R Oppenheimer. CERN. By courtesy of Emilio Segre Visual Archives.
9.1 P M S Blackett. By courtesy of the Imperial College of Science, Technology and Medicine.

9.2 F A Lindemann, Lord Cherwell. By courtesy of the University of Oxford.
9.3 Sir Henry Tizard. By courtesy of the Imperial War Museum.
10.1 Christiaan Huygens. Rijksmuseum. By courtesy of AIP Emilio Segre Visual Archive.
11.1 Particle shower in a cloud chamber. By courtesy of Sir Bernard Lovell.
11.3 Jets of high energy particles in nuclear emulsion. By courtesy of A C Breslin.

Chapter 1

INTRODUCTION

Why is there conflict? A simple answer is that a strong person or institution feels its interests threatened by a weaker person or institution. The reality is more complex. Usually the strong institution has an ideological, religious, scientific or political attitude which is threatened, or which it feels to be threatened. Thus conflict may be initiated without a desire for personal gain. Sometimes the threat may originate from a person who is not involved in the conflict, Copernicus and Darwin being examples. The battle over evolution was fought mainly between T H Huxley and Bishop Wilberforce, while Darwin stood on the sidelines motionless. We will return to this.

The term 'conflict' is a wide one. It may just mean contradiction or denial. It may mean a violent verbal interchange or physical violence on more or less equal terms. It may escalate to discrimination or even persecution, if one of the antagonists is much stronger than the other. The dictionary definition of persecution is: 'to subject to persistent ill treatment; to subject to penalties for heresy'. In common usage it means rather extreme repression, coercion or physical force.

The motives for initiating conflict vary widely. In most of the cases we will study they arise from a perceived threat to a religious, scientific or political orthodoxy, but obviously self-interest, direct or indirect, and the desire for power, may figure. There may be a threat to authority, there may be misunderstanding, or simply criminal violence. Arguments or quarrels may develop into violence. There may be a desire to divert attention from embarrassing circumstances, as in the case of Urban VIII, and his dalliance with the French against the interests of his nominal ally Spain. Religious and ideological struggles are hardly ever purely religious or ideological. Power is important also. Personality is a significant factor. Freud and Jung could hardly stay in the same room with each other in their later years, after an earlier close collaboration.

The present study will present brief treatments of some historical conflicts and their consequences in this chapter, and then move on in later chapters to cover some case studies in more detail: religion and science; internal to science; and politics and science. In order to introduce our case studies, this chapter will present some reflections on conflicts in the history of science.

Roger Bacon was born at Bisley in Gloucestershire, sometime between 1214 and 1220. He joined the newly formed Franciscan Order in about 1240. He coincided roughly with the origin of the Holy Inquisition in 1233, founded by Gregory IX.

Religiously he was quite orthodox, though having 'millenary tendencies'. His most important work was in optics, but he is celebrated for his development of gunpowder, and it was probably his chemistry that caused the trouble. It is not clear where he got his information. Gunpowder was invented by the Chinese, who, in what was probably the first international treaty on armaments, banned it in warfare. The ban held, until it was imported into the West.

It is not clear why Bacon ran into trouble with the authorities. The Order, when it imprisoned him, cited 'certain novelties' as the reason. No more is known [1]. Extraordinarily, he had been surreptitiously approached by the Pope, Clement IV (1265–8), who wished him to compile a compendium of 'all that was known'. Bacon was astonished. Even if he could carry out the enormous task, how could the material be smuggled out? He concluded that the Pope just did not appreciate how much was known, from the Greeks and the Arabs, to mediaeval scholars. The project lapsed, to Bacon's relief, in 1268, when Clement died, but this removed his only friend in high places. The Inquisition seems to have played no part in Bacon's imprisonment. It was purely an internal matter for the Order, and surprisingly, for although Bonaventure had systematized the Rule, the spirit of St Francis still remained to humanize. Bacon was imprisoned for several years, almost until his death in about 1293.

After Bacon, there were of course many trials of heretics, but none which involved a distinguished scientist. Copernicus, in the early sixteenth century, was left undisturbed; perhaps because he did not publish until almost his death. He was condemned by Luther and Melancthon, but the Roman Church did not get involved until the end of the century. The Inquisition, which had lapsed during the later Middle Ages, was revived and formalized in 1542, in Europe as a whole; somewhat earlier in Spain.

The new dispensation started with Giordano Bruno, in the 1590s. Suffice it to say that his case was complicated by the multiplicity of charges against him. Amongst his heresies he preached an infinite number of universes; and the not-yet-heretical Copernican theory. He was excommunicated by Catholics, Lutherans and Calvinists. Finally he was arrested by the Venetian Inquisition; reluctantly handed over to the Roman Inquisition, and tried and burnt by them in 1600. The Venetians would probably have left him be, but Rome demanded him. It was just a routine burning.

Galileo, of course was the *cause célèbre*, because of the highly placed people involved; his friendship with the Grand Duke, his standing as a scientist, his membership of the Academy and his international eminence. These will be considered at length in a later chapter. Initially the conflict was with a host of minor figures in the universities or the Collegio Romano, but ultimately it became a conflict with the Pope, Urban VIII, operating through the Roman Inquisition.

Thomas Kuhn says, not specifically of Galileo, but in general about the sixteenth and seventeenth centuries [2]:

> By now almost all surveys have come to include a chapter or major section on the Scientific Revolution of the sixteenth or seventeenth centuries. But those chapters often fail to recognise, much less to confront, the principal historiographic novelty which Butterfield discovered in the current specialists' literature and made available to a wider audience—the relatively minor role played by new experimental methods in the substantive changes of scientific theory during the Scientific Revolution. They are still dominated by old myths about the role of method...

This view must be modified when we think of Galileo's contribution to the Copernican theory. It is true that his telescopes were not vital to the scientific understanding of the heliocentric model—Kepler and his followers would ultimately have prevailed—but what Galileo did was to focus the attention of a generally educated but non-scientific public on the astonishing consequences of Copernicus' model. While this left Galileo himself vulnerable to his enemies, he had the majority on his side. Unfortunately this made him dangerous.

Moving outside the physical sciences and into modern times, we come, briefly, to the controversy over evolution. This is particularly interesting because the originator of the theory was not a prominent figure in the controversy. The main battle was fought between T H Huxley and Bishop Wilberforce, with Darwin more or less as an onlooker. Why was this? One reason, and probably the main one, was Darwin's health. He had extreme bouts of lassitude and intestinal problems, which continued for over twenty years. These have been dismissed by some biographers as hypochondria; Professor Medawar [3] considers the following probable. On 26 March 1835, in Argentina, Darwin was bitten by the Benchuca bug, the giant bug of the Pampas. On biting, this bug commonly injects a microorganism, *Trypanosoma cruzi*, into the bloodstream. It causes all the symptoms which Darwin possessed and often death at about age 40. In a sense, therefore, whilst Huxley did the fighting, Darwin the founder of the theory, was incapacitated. He has been unfairly criticized for his inactivity, as he was during his lifetime. We owe him a collective apology. In spite of his physical symptoms, Darwin wrote several books after *The Origin of Species* in 1859. We conclude that the true protagonists were Darwin and Bishop Wilberforce, with Huxley as a Sherpa.

Ludwig Boltzmann was Professor of Physics in Vienna, a most distinguished Chair. Nevertheless he was persecuted by a group of German chemists, led by Wilhelm Ostwald and Ernst Mach. They did not believe in atoms, but in energy levels. Only observable entities were allowed in the positivist scheme. Boltzmann was so hounded by them that he attempted suicide and, in 1906, succeeded. Three years later Ostwald admitted that atoms must exist; but it was too late for Boltzmann. Looking back it is hard to say what it was all about. Maxwell and Kelvin had demonstrated the existence of atoms and some of their important properties as early as 1870.

It is hard to imagine that Einstein in his middle years, with enormous world eminence, could get involved in conflict. The Nazis attempted it of course, but he was out of the country with his essential belongings and family and had many places to go. Nevertheless he was involved in quite violent argument. Bohr, Born, Margenau and all the others who took part would indignantly and sincerely deny any suggestion of persecution. We are his disciples, as Born said. Perhaps a family quarrel, with father outnumbered, is a better metaphor.

It was not of course persecution, but it was certainly conflict, though dressed up in the most polite language. Basically, Einstein, whilst accepting some of the theorems of quantum mechanics, did not believe that it was a complete theory, even in principle. There must be important hidden laws. This was contrary to the conventional or 'Copenhagen' school of theory. On the one hand was Einstein, with assistants such as Podolsky and Rosen, on the other a massive grouping with Bohr at the front, followed by Born, Margenau and others, virtually all the leading names of quantum theory except for Dirac, who remained aloof from the struggle. Schrödinger and de Broglie also had doubts about the Copenhagen interpretation, but did not get deeply involved. They represented the older quantum theory, or wave mechanics.

In Russia there was also an orthodoxy, and a very rigorous one. Marxist theory abhorred idealism, and quantum theory was banned under Stalin. Most of the theoretical physicists got away with a rebuke, but some disappeared. More serious was the dispute between Vavilov and Lysenko over plant genetics. Vavilov was a man of great international reputation, a foreign member of the Royal Society, an honour which is not given lightly. Lysenko was a peasant in origin, and had the ability to persuade and convince peasants. It was this which made him indispensable to Stalin, who had just collectivized the land. Miracles were needed, and Lysenko supplied them. By a process of pasteurization called vernalization, he claimed to convert winter wheat into spring wheat. Vavilov challenged this on genetic grounds, and promptly disappeared. Was the aggressor Stalin or Lysenko? It hardly matters. Lysenko was the faithful servant, the Fra Maculano.

Shortly after, in 1953, came the Oppenheimer case. Oppenheimer had communist connections, his brother, his wife and his mistress had all been members of the party in the 1930s. Oppenheimer himself was never a member, but maintained ambivalent connections, even when director of Los Alamos. After the war he left to become director of the Princeton Institute for Advanced Studies, but maintained important Atomic Energy Commission (AEC) committee memberships. At the instigation of General Nichols, Groves' successor as general manager of the AEC, a security check was made on Oppenheimer, followed by a full-scale enquiry. He was declared a security risk and returned to Princeton, causing a major battle among America's scientists. Their sympathies were overwhelmingly with him, but there were rather few public protests. There he continued as director until he was rehabilitated by President Kennedy, and given the Fermi Medal. However, since his principal antagonist at the trial, Edward Teller, had been awarded it the year before, it was poor consolation.

It is difficult, with these short summaries, to draw conclusions that apply to all, but one thing is clear. The scientist is not a revolutionary trying to overthrow the established order. We will return to this point.

In two of our detailed case studies, Galileo and Oppenheimer, the parallels are remarkable. Both achieved high responsibility at an early age, and a high degree of public fame in their forties. They were protégés of very powerful people. There were conflicts but they did not seem important. Then suddenly, out of the blue, when they were elder statesmen loaded with honours, the blow struck. They were humiliated and disgraced. Once that was done, the penalty was rather mild.

Even in the Einstein case there are similarities. To his own generation he was a demi-god, but to the younger generation of quantum physicists he was a slightly pathetic old man, and this change in opinion also happened rather quickly, though not as quickly as in the other two cases.

In the sixth case study we will trace the development of the atomic hypothesis through the centuries. This is dangerous, because it is very easy to impose a spurious continuity on a subject. Kuhn says [4]:

> Scientist-historians and those who followed their lead characteristically imposed contemporary scientific categories, concepts, and standards on the past. Sometimes a specialty which they traced from antiquity had not existed as a recognised subject for study until a generation before they wrote. Nevertheless, knowing what belonged to it, they retrieved the current contents of the specialty from past texts of a variety of heterogeneous fields, not noticing that the tradition they constructed in the process had never existed.

This limitation would not seem to affect the controversy outlined in Chapter 5, since nearly all the relevant components of the quarrel originated in one generation. Material before that was common ground. Kuhn does not deal, in *The Essential Tension*, with conflict, but generally the material presented here is compatible with his view of science. For more recent controversies, such as the Einstein, Podolsky and Rosen affair, there might be some difficulty in applying Kuhn, since he does not recognize the distinction between *basic* and *fundamental* research, at least in these early essays. Prior to about 1900, most basic research in physics could be regarded as fundamental, since many totally new principles were appearing. Now, fundamental research is largely limited to cosmology and elementary particle physics. By fundamental we mean directed to new physical principles; basic research does not involve new principles, but is not applied or commercial.

In the three chapters which follow this, we describe five conflicts between individuals and the churches. This may be misleading. It does not imply that the churches, whether Eastern, Catholic, Calvinist or Lutheran, were enemies of science. The Church was in fact for 1000 years the repository of knowledge, secular as well as sacred, and often was forced into a defensive position on secular issues. This does not of course excuse St Cyril's followers in their murder of Hypatia, or the Franciscan Order in their treatment of Roger Bacon, the Roman Inquisition in its treatment of Giordano Bruno, the same Inquisition in its treatment of Galileo, or both Lutheran and Catholic treatment of Kepler. Hypatia was a threat because she was more intelligent than St Cyril, as Roger Bacon was more intelligent than his Franciscan confrères. Giordano Bruno was executed on religious grounds, certainly; his defence of Copernicus, while important historically, did not figure explicitly at his trial. The condemnation of Galileo was a complex affair which deserves a whole chapter, but had more to do with jealousy than with religion or science. Kepler's misfortunes arose in large part from what would now be called ethnic cleansing. Few people blame the churches today for the situation in Bosnia.

The problem is one of authority, not restricted to religious authority, and it is of course worst when authority is insecure. A secure authority can afford tolerance. In the fourth and fifth centuries the Church, while holding imperial approval, was riven with conflict; and the ancient, pagan, Greek culture was seen as a threat, even though it was almost dead. Similarly in the sixteenth and seventeenth centuries the Reformation and later the Thirty Years War, threatened a complete breakdown. Galileo was not a threat to the Bible, he was not even a serious threat to Aristotle, he was a threat to the security of Urban VIII. The parallel may be extended into the twentieth century. Whether domestic communists were really a threat to the security of the United States can be argued. It seems fairly certain that Oppenheimer was not a threat, but was caught up in anti-communist hysteria.

In the internal conflicts, the situation was somewhat different, although the desire to be proved right and hence to be dominant was obviously an element. Bohr did not want to defeat Einstein, but he wanted to be proved right. This is not necessarily harmful, indeed it is probably necessary for any scientific progress at all. It does, however, affect the interplay of personalities; in the Boltzmann case hostile, in Bohr and Einstein, friendly. The N-ray conflict was a tragic interaction between well-meaning participants. Blackett, Tizard and Cherwell also took part in a tragic confrontation.

Non-scientists are often surprised by the degree of seriousness and dedication amongst scientists. At least three suicides of distinguished scientists in this century can be traced directly to situations in physics. Amongst composers or painters, it does not surprise, but in science it does. The degree of single-mindedness involved, and the extent to which one's whole world is bound up with science, is not appreciated.

Chapter 2

RELIGION RAMPANT: HYPATIA, ROGER BACON, GIORDANO BRUNO

Religion and science should be symbiotic, and sometimes are, but usually there is mistrust and suspicion between them. It is not clear why this should be so, since they have objectives in common, but throughout most of recorded history, it has been so, except in the very early days when scientists were priests. We will not attempt an analysis of the problem in this chapter, but simply record three instances of persecution of scientists by the religious. A fourth, the Galileo case, will be treated in a later chapter. It was a relatively gentlemanly affair. These three were not.

2.1. Hypatia of Alexandria (c370–415 AD)

She was the greatest woman scientist of antiquity of whom we have a record. M Alic has said that she was the most eminent woman scientist before Marie Curie, and other writers including Gibbon, have written in similar terms. She was renowned for her erudition, beauty and virtue, but these led to her murder by a demented mob. We will return to this, but first we must say something about Greek science and its origins, for this was the material which she, coming at the end of the classical period, inherited. The centre of Greek philosophy was Athens, but Alexandria was the crown of Greek science. Founded by Alexander the Great in the flatlands of the Nile delta, in 323 BC, the year of his death, by 300 BC it had attracted scholars from all over the known world. The Museum, or university, lasted a thousand years, until the final destruction of its great library by the Mohammedans in 642 AD.

Before Alexandria, of course, there was scientific and technical activity back to the earliest times. Its scientific component was mainly astronomical, with the very practical aims of regulating the planting of crops, predicting the flooding of the Nile and so forth. With this practical aim went speculation about the origin and nature of the universe. At first it was mythological, though reaching high levels of numerical accuracy. The oldest activity that can reasonably be termed scientific (as distinct from technical) was that of the Sumerians in the fourth millennium BC. Their influence in Egypt caused the determination of the length of the year, to a high degree of accuracy. The Egyptians had taken the year to be exactly 365 days—twelve months of thirty days and five sacred days. They found over many years that the civil year got out of step with such events as the flooding of the Nile, and the planting of crops. Only a long-lived and stable civilization, of course, would have been able to detect these discrepancies, since they took hundreds of years to show up. Looking for a clock more precise than the day, they found it in astronomy, the rising of the stars. They chose to observe the moment at which Sirius became visible in the early morning dawn (heliacal rising). This advanced through the civil year at the rate of one day every four years, so the year was set at $365\frac{1}{4}$ days, the quarter days being accommodated in a leap year every four years. For civil purposes, however, they retained the year of 365 days exactly. Later Julius Caesar introduced the Julian year, which had leap years, and lasted until the sixteenth century AD.

Another more dubious reason for being interested in the stars was astrology. It should not be dismissed out of hand in the history of science. It provided a living for Tycho Brahe and Kepler, among others, and from the earliest times promoted a close study of the stars. It was practised in Babylonia in about 3000 BC, giving rise to accurate measurements of planetary motion.

The Babylonian tradition lasted for 3000 years of accurate observation, aided by a simple and powerful system of positional arithmetic (based on a modulus of 60, not 10 as in our modern system). In the sixth century BC they discovered the Saronic cycle of 233 months, which governs eclipses. They made amazingly accurate measures of other astronomical periods, particularly that of the lunar month:

Nabarriannu (about 500 BC)	29.530614 days
Kidinnu (about 383 BC)	29.530594 days
True value	29.530596 days.

Kidinnu's value is accurate to about a fifth of a second. They were also interested in commercial transactions and prepared tables of compound interest (rates varied from 20–30%). Another important factor in the development of science was the use of the stars for navigational purposes at sea. This was developed particularly by the Phoenicians.

2.1.1. *Thales and Anaximander*

From the sixth century BC onwards, much of the information was transferred to the Greek city-states, particularly in Ionia. The first philosopher of note here was Thales of Miletus (c623–c564 BC). Herodotus claims that he was in fact of Phoenician extraction, but this is doubtful. Certainly though, Phoenician influences were strong in his thought. T L Heath says of him: 'Statesman, engineer, man of business, philosopher, mathematician and astronomer, he covered almost the whole field of human thought and activity'. He travelled in Egypt, where he studied for a while, and in Babylonia, and doubtless picked up extensive astronomical information. He must also have acquired expertise in geometry, particularly in Egypt, which was advanced in this area.

He is said to have predicted an eclipse which occurred during a battle, the darkness being so complete that fighting had to stop. A peace ensued because of this direct intervention of the gods. Thales produced the first scientific model of the universe. While it bore little relation to reality, it was free from mythological elements.

A fellow Ionian and contemporary was Anaximander (611–547 BC). He wrote a book on geometry, but was particularly interested in astronomy and geography. He had an evolutionary cosmology, of some biological interest. He introduced the sundial into Greece from Babylonia. When he died the school at Miletus became more interested in philosophy, and the scientific work died away by about 400 BC.

2.1.2. Pythagoras

The Pythagorean school was of great importance, but little is known of its founder. He was born in Samos in Ionia, and like Thales is said to have come from a Phoenician background. He is known to have left Samos in 530 BC, whence he settled in Italy. He founded a kind of secret brotherhood, whose tenets are still not entirely known. The search for knowledge was its basic aim, and knowledge when found, could be communicated only to one other person. Harmonies, as in music, were important and numbers were all believed to be rational. The person who discovered the irrationality of the square root of 2 is said to have been taken out to sea and drowned.

The theorem of Pythagoras is the most important achievement of the school, though the Indians and Babylonians knew examples of it earlier. The preoccupation with the aesthetic properties of certain numbers influenced later Greek schools, in spite of the secretive character of the brotherhood. They believed that space too was composed of a chain of integral finite numbers, and the irrationality of certain numbers was ultimately fatal to their cosmology. In astronomy some Pythagoreans were ahead of their time, in holding that the Earth moves and rotates about a central fire, though this central fire was never identified with the Sun. The model was not generally accepted.

2.1.3. *Plato and Aristotle*

The greatest period in Greek philosophy was that from about 430–320 BC, and was dominated by Plato and Aristotle. While Aristotle was a great observational biologist, and Plato stressed the importance of mathematics, it was not a good time for science. Many erroneous views were held, and because of the influence of the two great philosophers, their views persisted for two thousand years. Thus the Aristotelian universe of concentric spheres was still held to be true in seventeenth-century Europe, and the crystal spheres were shattered only by Copernicus, Kepler and Galileo. Some later Pythagoreans like Ecphantus and Heraclides introduced a universe in which the Earth rotated on its axis every day, and in which some planets, but not the Earth, revolved around the Sun. Again, these ideas were not accepted.

2.1.4. *Alexandria*

Greek science took a great leap forward with the foundation of the Museum of Alexandria. On the death of Alexander the Great in 323 BC, his kingdom was divided, and Egypt was taken by Ptolemy, who chose Alexandria, still unfinished, as his capital. The Museum was a university with a strong research element. It lasted for a thousand years, and at its highest point had a library with 600 000 manuscripts in many languages. Many scholars migrated there from Athens and other centres, leading to a massive concentration of talent. In mathematics the greatest names were Euclid, Archimedes (who was also a great physicist) and Apollonius who also worked in astronomy. Hero, Diophantus and Pappus came later.

In astronomy Aristarchus was perhaps the greatest, indeed one of the greatest of all time, but Eratosthenes, Hipparchus and Ptolemy were outstanding also. Both Hipparchus and Ptolemy used information from Apollonius' work, the most important of which was on the conic sections. Apollonius' work is also significant because it was among those edited by Hypatia and thereby preserved. Similarly her edition of Ptolemy is essentially the one we have today. She also wrote a commentary on Diophantus. We will restrict our discussion to astronomy.

2.1.5. *Aristarchus (c310–230 BC)*

Not much is known about the life of Aristarchus, but he was born in Samos. His most important work was 'On the sizes and distances of the Sun and Moon', which is still extant. It is a genuine scientific and observational treatise, far ahead of its time, and was the basis for estimates of the size of the solar system for many centuries. The principle involved is quite simple. It is recognized that the Sun is much further away from us than is the Moon. At the moment of half-moon (figure 2.1) the angle MES, will be nearly, but not quite, a right angle. The angle EMS is exactly a right angle. By measuring the angle MES we can determine the shape of the triangle and hence the relative lengths of

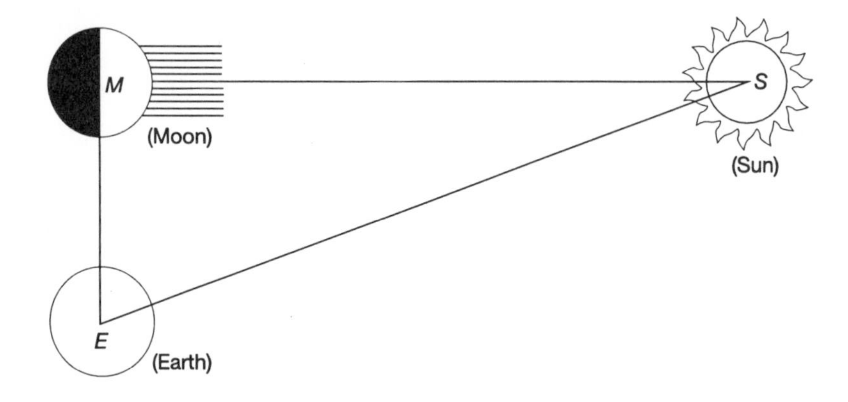

Figure 2.1. *Aristarchus' model of the solar system.*

ME (the Moon–Earth distance) and SE (the Sun–Earth distance). The difficulty arises because the angle MES is very nearly a right angle, and the accuracy of SE depends not simply on the value of MES, but on its difference from a right angle. Also the exact moment of half-moon is difficult to determine.

Aristarchus measured the angle MES to be 87°. In fact the true value is 89°51′, so that the true distance is about twenty times greater than his value. In order to find a numerical value of ES we must know the distance ME from the Moon to the Earth. To do this Aristarchus used the size of the shadow cast by the Earth on the Moon during a partial eclipse, relative to the size of the Moon. He estimated the ratio at about 7. In fact it is 4, but this error is not too serious. Now, if we know the size of the Earth, and this was known approximately, and we measure the angular size of the Moon seen from the Earth, we can determine ME. Knowing this Moon–Earth distance we can calculate the Sun–Earth distance SE. The value which Aristarchus found was about three million miles. The true value is about thirty times greater, ninety million miles, but this error is less important than the realization that the solar system is very large.

Aristarchus' value was accepted for two thousand years, until the late seventeenth century AD. An inspection of the angular size of the Sun now tells us that the Sun is far larger than either the Earth or the Moon. Is it reasonable then, to suppose that it rotates about the Earth? Aristarchus said no, the Moon rotates about the Earth, but the Earth and Moon rotate about the Sun, with the other planets. His achievement was truly epoch-making, it demonstrated the magnitude and geometry of the solar system. This was the highest point of Alexandrian astronomy, scientific in concept and performance. He was also aware of the problem of stellar parallax, the variation of stellar positions when the Earth moves, and the answer to it. Unfortunately it was not widely accepted. An

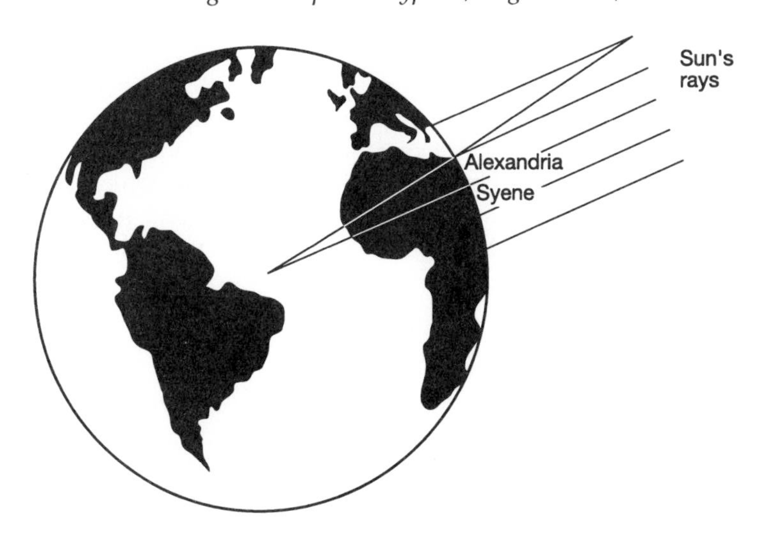

Figure 2.2. *Eratosthenes' model of the Earth.*

improvement in the measurement of the Earth was made by Eratosthenes about fifty years later.

2.1.6. Eratosthenes (c276–195 BC)

He was a very important figure in the Museum, being chief curator of the library. He improved an existing method of measuring the circumference of the Earth by using a fact which he knew by hearsay. It was said that on midsummer's day at Syene (now Aswan) the Sun was directly overhead, so that the bottom of a vertical well was illuminated (that is, Syene lay on the tropic). At Alexandria on that day at noon the Sun was 7°12′ below the zenith. Assuming that Syene was due south of Alexandria, and measuring the distance between them (according to legend this was done by marching a platoon of soldiers between them and counting the number of paces taken), the size of the Earth can be calculated (figure 2.2).

Eratosthenes concluded that the circumference of the Earth was 252 000 stadia. According to Archimedes it had previously been taken as 300 000 stadia, not a very large difference. Unfortunately we do not know the exact length of the Egyptian stadium. It is believed to have been 517 ft, slightly different from the Greek stadium. Taking that value, we find a circumference of 24 650 miles; the true value being 24 875 miles. The agreement is remarkable and must be fortuitous, since there were several approximations in the calculations, and Aswan is not quite on the tropic. Nevertheless this measurement put Aristarchus' views on a firm footing. Eratosthenes also measured the 'obliquity of the ecliptic', the tilt of the Earth's axis. He found a value very close to the correct value.

2.1.7. Hipparchus (c190–120 BC)

He discovered the precession of the equinoxes, a slow movement of the Earth's axis, 25 800 years being required for a complete cycle. He also made very accurate measurements of several astronomical constants. His catalogue of star brightnesses was the basis on which the stellar magnitude system was defined in the nineteenth century AD (astronomy is a very conservative subject). In mathematics he made advances in spherical geometry.

2.1.8. Claudius Ptolemy (c100–151 AD)

He had no connection with the ruling family of Egypt. He appeared after a gap of two centuries, in which Alexandrian astronomy was static. Objectively he wás not the greatest Greek astronomer, but he was by far the most influential for more than a thousand years. Also he was important for our present topic since Hypatia's standard edition of his greatest work, *The Almagest*, preserved a text which was vital to astronomy until Copernicus.

The Almagest contains thirteen books, some mathematical and some astronomical. Book 1 is on trigonometry and includes a calculation of the value of π. Ptolemy gives a value of 3.14167...; the true value is 3.14159.... This indicates the accuracy which some of his calculations achieved. Two other books contain the places of about a thousand stars, following Hipparchus closely. But his most influential work is on the motion of the planets. Much of it is based on his own observations. Some years ago the world of astronomy was shocked by the charge that some of his measurements were fraudulent, fudged to make the answers come out as desired. He was defended against this accusation in the exalted pages of the *Quarterly Journal of the Royal Astronomical Society*, and the controversy raged for some time, but ended indecisively. Whatever the truth there is no doubt that nearly all of Ptolemy's observations were genuine. Copernicus could not have worked without him. But of course his model of the universe was wrong (figure 3.3) and unnecessarily complicated. Some previous Greek astronomers, as we have seen, put Mercury and Venus in direct rotation about the Sun, which simplified their orbits considerably. Ptolemy rejected this and made them, with the outer planets and the Moon, revolve around a fixed Earth. To obtain agreement with observation, all the planets required two circles to describe their motion: a large circle called the deferent and a smaller circle with its centre moving on the deferent. This was called an epicycle, and carried the planet. The scheme explained the observed forward and retrograde motions of the planets, being particularly successful in the case of Mars. With time and improved observations, however, the number of epicycles had to be increased to eighty. (Even Copernicus needed thirty-four epicycles to fit the observations to his system. The reason for these extra circles is that the orbits are in fact ellipses and both Ptolemy and Copernicus were trying to fit to circles.)

The nineteenth-century mathematician de Morgan said that the adoption of Ptolemy's system was fortunate, because the culture was not ready for Copernicus; perhaps this sociological view is valid in science generally. Nevertheless it held up progress, by making calculation more difficult than it need have been. Ptolemy also wrote on optics, and understood refraction, both in terrestrial and astronomical situations. He made maps, but these were based mainly on hearsay.

After Ptolemy, Alexandrian science went into decline. Control of Egypt was in the hands of Rome. The Romans tolerated science but were not really interested. In the fourth century AD there was a brief renaissance. For the first time in antiquity, science was led by a woman.

2.1.9. Hypatia's life

As the leading woman astronomer of her time, and perhaps of all antiquity, and in the circumstances of her death, Hypatia has become somewhat of a cult figure. Novels and plays have been written about her in several languages, and there are two journals named after her. She appears on the World Wide Web.

The date of her birth is uncertain, but she was murdered in 415 AD. Charles Kingsley, in his romantic novel *Hypatia*, puts her at about twenty-five then, which would imply a date of birth of about 390. This is undoubtedly too late. 370 is frequently given in encyclopaedias, but this too is late; Synesius, an important official and afterwards a bishop, studied under her in the period 390–395. It seems unlikely that she would have been ready to take on students of such eminence at the age of twenty to twenty-five. Dzielska [1] makes her birth date 355, which would make her 60 at the date of her murder. She bases this on the age and standing of Hypatia's students, particularly Synesius. Thirty-five to forty is a reasonable age at which to have high standing in the community.

She was the daughter of an astronomer and astrologer named Theon, who was born around 335 AD, and probably died around 400 AD. He was also skilled in physics and mathematics, writing a student edition of Euclid's *Elements*, as well as the *Data* and the *Optics*. Modern editions of these derive from them, so that Theon has influenced generations of schoolchildren. He also wrote a commentary on *The Almagest* (which was followed by his daughter's standard edition of the work).

Hypatia was a greater mathematician than her father, but like him she was not a profoundly original thinker like Aristarchus or Euclid. She was a commentator and above all a patient and accurate editor. Several major Greek works, above all *The Almagest*, owe their survival to her. Her commentary on the works of Apollonius was also important. Since Copernicus could not have worked without *The Almagest*, Hypatia must be regarded as a significant contributor to modern astronomy. She was also a charismatic and erudite teacher, in philosophy as well as in mathematics and science. Her students were devoted to her, and we owe much of our present knowledge of her to Synesius. He wrote about 150 letters that have survived.

She had a very strong character. Dzielska says [2]:

> She certainly was not endowed with an enticing, pleasing, or sympathetic character. Such qualities do not fit her. One may say of Hypatia—and the sources do—that she was endowed with uncommon force of character and ethical fortitude.

Over the years, assessment of her character has varied. In 1720, John Toland published a pamphlet entitled:

> Hypatia or, the history of a Most Beautiful, Most Virtuous, Most Learned and in every way Accomplished Lady; Who Was Torn to Pieces by the Clergy of Alexandria, to Gratify the Pride, Emulation and Cruelty of the Archbishop, Commonly but Undeservedly Titled St Cyril.

This was rapidly answered by Thomas Lewis:

> The History of Hypatia, a Most Impudent School-Mistress of Alexandria. In Defence of Saint Cyril and the Alexandrian Clergy from the Aspersions of Mr Toland.

Edward Gibbon takes a very similar view to Toland in his *Decline and Fall of the Roman Empire* (vol 5, p 117, Bury Edition, 1909), as might be expected from his central thesis that classical civilization was destroyed by the rise of Christianity:

> He [Cyril] soon prompted, or accepted, the sacrifice of a virgin who professed the religion of the Greeks . . . Hypatia, the daughter of Theon the mathematician, was initiated in her father's studies; her learned comments have elucidated the geometry of Apollonius and Diophantus, and she publicly taught, both in Athens and Alexandria, the philosophy of Plato and Aristotle. In the bloom of beauty, and in the maturity of wisdom, the modest maid refused her lovers and instructed her disciples; the persons most illustrious for their rank or merit were impatient to visit the female philosopher; and Cyril beheld with a jealous eye the gorgeous train of horses and slaves who crowded the door of her academy. A rumour was spread among the Christians that the daughter of Theon was the only obstacle to the reconciliation of the prefect and the archbishop; and that obstacle was speedily removed. On a fatal day, in the holy season of Lent, Hypatia was torn from her chariot, stripped naked, dragged to the church, and inhumanly butchered by the hands of Peter the reader and a troop of savage

> and merciless fanatics: her flesh was scraped from her bones with sharp oyster-shells, and her quivering limbs were delivered to the flames. The just process of inquiry and punishment was stopped by seasonable gifts; but the murder of Hypatia has imprinted an indelible stain on the character and religion of Cyril of Alexandria.

Contemporary sources discovered since Gibbon's time (mid-eighteenth century) say that the implements used were not oyster-shells, but sharp fragments of pottery. It is also inferred that she was raped. Gibbon does not answer two crucial questions: who murdered Hypatia, and why?

The first of these questions has been answered. She was murdered by a group of Parabolans. These were a paramilitary militia employed by individual churches to defend their interests at a time when theological conflicts were particularly violent. They also had some medical duties. It is assumed that this particular group had some connection with Cyril. Probably some monks were present but it seems unlikely that they took any close part in her murder.

Why was she murdered? This is more difficult to answer. There are a number of possibilities.

(i) The archbishop Cyril saw her as a threat to the power of the church in Alexandria. She was friendly with the Prefect, Orestes, who was in charge of the administration of the city. Orestes was on bad terms with Cyril, and had been attacked by monks who slightly injured him.

(ii) Although she was generally regarded as a pagan, she had close links with Christians. One of her students, the faithful Synesius, became a bishop, other Christians were powerful politicians at the imperial court, and large landowners. All of these were seen as a threat by Cyril.

(iii) In a propaganda campaign, Hypatia's academic activities were presented as witchcraft. It was recalled that her father had practised astrology. It is unlikely that the witchcraft claim was believed by those who started the rumour, but it would be accepted by the Parabolans, who were notoriously superstitious.

(iv) Bitter theological controversies were going on, particularly between Cyril and Nestorius, the patriarch of Constantinople. Cyril had already evicted the Jews from Alexandria. Possibly Hypatia was suspected of connections with one or more of these groups.

(v) Her science was seen as a threat to orthodox Christianity. This is the view which was taken by several commentators in the nineteenth century and since. Her scientific and mathematical achievements have, perhaps, been slightly exaggerated. Dzielska says [3]:

> The same account and accusations persist today in historical studies of Hypatia, in various kinds of dictionaries and

encyclopedias, in histories of mathematics, and in works dealing with women's contributions to the history of science and philosophy. Thus the *Dictionary of Scientific Biography* (1972) characterises her as 'the first woman in history to have lectured and written critical works on the most advanced mathematics of her day'. A W Richeson, writing about the 'celebrated mathematician-philosopher Hypatia', asserts that after her death 'we have no other mathematician of importance until late in the Middle Ages'. Similarly R Jacobacci states that 'with her passing there was no other woman mathematician of importance until the eighteenth century'. M Alic describes Hypatia as the most eminent woman scientist before Marie Curie. B L Van der Waerden reiterates the theme that Alexandrian science ceased with her death.

The perception of science as a threat to religion has at times been put forward by both friends and enemies of religion. It was commonplace in the eighteenth century among such writers as Voltaire. When combined with Neo-Platonic philosophy as in the case of Hypatia, it is easy to see that Cyril would have seen her as a threat.

(vi) She was a woman, competing with men. This could not be tolerated.

Altogether, there were many reasons for wanting her out of the way. It has not been suggested that Cyril was present at her murder, but he must have been in some way involved. The appalling brutality of her murder, and the fact that her body was burned, suggest that the assassins saw her as a witch.

2.1.10. Conclusions

She can reasonably be described as a scientist for she had a profound and accurate knowledge of the major astronomical and mathematical works available. She was not a major original figure, but very few working scientists are. She made very valuable contributions in the editing and preservation of important works. While undoubtedly there were important religious factors involved in her murder, there was sufficient astronomy to warrant regarding her as a martyr for science.

St Cyril went on from strength to strength, though his powers were briefly restricted after Hypatia's death, and the number of Parabolans was reduced. At the Council of Ephesus in 431 AD, he completely defeated Nestorius. The latter's followers had to flee, first to Mesopotamia, then to Persia. There they settled and some remain today. Jeans [4] says:

> Here they were free to occupy themselves with literature and science, writing original works in Syriac, their own native

language, which had now become the common language of Western Asia, and translating the works of Aristotle, Plato, Euclid, Archimedes, Hero, Ptolemy and many others into the same tongue.

The final end of the Alexandrian school came in 642 AD, when the Mohammedans conquered the city, and destroyed the remainder of the great library. The Caliph Omar is said to have justified this final act of vandalism on the ground that: 'If these writings of the Greeks agree with the book of God, they are useless and need not be preserved; if they disagree, they are pernicious, and ought to be destroyed'. Abulpharagius records that the volumes kept the four thousand baths of the city in fuel for six months—an obvious exaggeration, for even if there were 400 000 volumes left in the library, the average fuel ration per bath would be only four volumes per week.

When the Arab Empire became consolidated, more liberal views of Greek learning prevailed. The manuscripts were translated from Syriac into Arabic, and finally, in the twelfth and thirteenth centuries, into Latin. So they came back into the West, and triggered an intellectual renaissance which led to Copernicus, Galileo and Newton. Thus Hypatia's life, work and death were not wasted.

2.2. Roger Bacon and the Franciscans

Roger Bacon was the greatest physicist produced by Oxford in nearly eight centuries. By and large the Isis and the Cherwell have harboured the quieter humane activities. This was certainly not true when Roger Bacon was there as he is best known for his work on explosives. He was born of a well-to-do family in either Ilchester in Somerset or in Bisley in Gloucestershire. The family had been wealthy, but had lost a great deal during the reign of Henry III (whom Bacon met on at least one occasion). Roger Bacon was an extrovert, who mastered the classics and the quadrivium (geometry, arithmetic, music and astronomy) early in life. Like Bruno after him, he had a prodigious memory, and the speed with which he could produce books was enormous, in spite of the restrictions placed on him by the Franciscans for much of his life.

His date of birth is uncertain; earlier authors put it at 1214, but more recently this has been revised up to 1220. His date of death is similarly uncertain, but he was alive in 1292. He was, therefore a contemporary of Thomas Aquinas, and of Bonaventure, and was born at the height of the Aristotelian renaissance and at the same time as the Fourth Lateran Council. All these points will be mentioned again. This was a period of enormous activity. Heresies abounded and were crushed; the scribes scribbled madly to keep up with the flood of Greek books from the

Arabs, and in the middle of all this Thomas Aquinas constructed his great synthesis which was to dominate the Catholic Church's teaching for seven centuries. An English Dominican, Robert Bacon, may have been Roger's uncle (he is recorded as lecturing at Oxford, and as being influential with such people as Robert Grosseteste, a future Chancellor of the University, and bishop). After Oxford, Roger Bacon went to Paris, and attended the university there, probably taking his Master's degree in Paris rather than at Oxford. He was there in 1244, for he claimed to have heard Alexander of Hales and William of Auvergne, both of whom died before 1250.

Bacon lectured in Paris before 1250. He was an Aristotelian, but with Neo-Platonic influences. The shadow of Averroes, condemned by the Fourth Lateran Council, raises its head. In spite of Bacon's family's considerable losses under Henry III, he still had resources, and he spent much of them after 1247 on books, experimental equipment and the training of assistants. He became obsessed with science, spending the enormous sum of two thousand pounds, according to his own reckoning, on instruments, materials and assistants, thereby raising the suspicions of his superiors, for science was alchemy, close to magic, and magic was close to witchcraft.

The problem surfaced after 1250 when he returned to Oxford. It is not known when he joined the Franciscans; a somewhat surprising decision. They were not a very intellectual Order, and went into the universities only to be better equipped to combat heresy. It was far from their founder's intention. Sometime before 1257 the Oxford Franciscans began to have qualms about him, and decided to send him back to Paris, away from the dangerously scientific air of Oxford. Under Bonaventure the character of the Order was changing, and the first thing they did to poor Bacon was to place him under restraint. For ten years he was under more or less solitary confinement, without books, paper or instruments. In spite of this his reputation as 'Doctor Mirabilis' spread throughout Europe. But then fate intervened: Guy de Foulkes, who had been papal legate in England in 1264–5, and had made repeated efforts to contact Bacon; became Pope under the title of Clement IV.

The Pope communicated secretly with Bacon, demanding from him a treatise which would cover all the sciences; but it must be compiled without the knowledge of his superiors. Bacon, nonplussed, misunderstood, and visualized a monstrous encyclopaedia compiled by many authors. Apart from writing it, the logistic problems of smuggling it out to the Pope would be enormous. The problem probably arose partly because the Pope did not appreciate how much was known.

Nevertheless, Bacon embarked upon the work. He wrote later in the *Opus Tertium* that it was with great joy that he embarked upon this task, even though it must be secret. His first work, completed in less than two years, was the *Opus Majus*. Basically it was a propagandist

Figure 2.3. *Roger Bacon (1214–1294), scholar, sends his manuscript secretly to the Pope from prison. Reproduced by permission of the Mary Evans Picture Library.*

work as well as an encyclopaedia, demanding that the sciences should be taught in the universities. He then embarked upon an abridged treatment which became called the *Opus Minor* or *Opus Secundum*. Finally the *Opus Tertium*, 75 chapters long, is clearly the beginning of a much larger work. The *Dictionary of National Biography* describes the beginning of the book [5]:

> The opening chapters of the writing called *Opus Tertium* give a very vivid picture of Bacon's circumstances when he received this mandate, of the joy with which he hailed the opportunity afforded to him, of the manifold difficulties in the way of completing the work on which he forthwith entered, and of the plan he adopted for laying the substance of his reflections before his friendly auditor.

Whether Clement actually helped to obtain more freedom for Bacon is not clear; perhaps enough to help him to get materials and books. Certainly by 1267 he was more or less free. There is no evidence that he was subjected to physical violence or even interrogation during his years of imprisonment. Clement died in 1268, after only three years as Pope, but Bacon remained, again, more or less free. He was not to remain so. In 1277 he was imprisoned again, and remained in prison for two years. The reason was 'suspected novelties' in his teaching [6].

2.2.1. Philosophy, theology and science in Bacon's teaching

These three subjects are intertwined, which was part of the problem of those who had to deal with Bacon. One suspects that they were not very sophisticated. The heresy hunters were their colleagues the Dominicans, but for reasons of pride they did not want to call them in. What was the problem, anyway? Today, Bacon would strike us as credulous to the *n*th degree [7], obsessed with the millennium, a believer in astrology, alchemy, and moreover not very easy to get on with. The inhabitant of the next cell in the cloister must have found it very difficult to live with exploding gunpowder of various qualities, with complicated optical experiments, and attempts to make airships. It is understandable that they wanted to send him off to Paris. The Franciscans had become harder, but they retained something of the gentle spirit of their founder. In spite of all these aberrations, Bacon was a great physicist, although everything that he wrote, no matter how scientific it purported to be, was mixed up with philosophy and theology, as well as the occult sciences. His important contribution to science was his repeated insistence on experiment. There was to be nothing like this until Galileo.

However, Bacon's understanding of experiment was very different from Galileo's. He was closer to a modern chemist than to a modern physicist, in spite of his work on optics. His work on optics is an example of very good science. He put forward the laws of refraction, reflection and spherical aberration, accurately, in spite of mathematical limitations. He appreciated that the velocity of light was very high, but he did not move further to the variation of that velocity with the medium of transmission. It is this mixture of sophistication with credulity that makes Bacon so fascinating. In some ways he was sophisticated far beyond his century; in others, surprisingly naive. In some ways he almost reached the science of the seventeenth century; in others he was primitive.

His theology and philosophy were interesting apart from their scientific connections. They were probably largely responsible for his troubles. At a time when theology was being codified under Aristotle, he, though an Aristotelian, was calling for more use of Scripture in theology. He also, and this is curious in view of his interests, was calling for a reduction of philosophy in the study of religion. It would be replaced by languages, mathematics and practical physical sciences. He disapproved of scholastic theology, thereby setting him against the Dominicans. Rather, he held that simple doctrine was sufficient for salvation through preaching. Thus he appears as a reformer before his time—three centuries before his time.

Consequently there was plenty of theological ground for preventing him from writing; but it was not in fact the reason given. They were frightened of his science. Theologically the Order was advanced and without the subtle convolutions of the Dominicans. It was unfortunate that Bacon missed, by a few years, the advent of Duns Scotus, who cast a blinding light on mediaeval philosophy, though they would not have got on. Duns Scotus entered the Order a few years before Bacon's death, mixing with much the same people in Oxford and Paris. Who knows what they could have done together?

Bacon's science came from a variety of sources. His most famous import from the East, gunpowder, came originally from China. He seems to have been aware of its potential in warfare, but did not develop it. Here he followed the Chinese, who, nearly a thousand years before, had, after a summit conference, decided that the use of gunpowder for weapons was inhumane and must be prohibited. Surprisingly, the agreement held.

Bacon's failure to develop gunpowder for military use appears to have arisen from humanitarian grounds, and from appreciation of its potential. It was developed in the next century.

After Bacon's comparative freedom during the reign of Clement IV and for ten years after, the blow fell again, and again from his fellow Franciscans. The *Encyclopaedia Britannica* says: 'Sometime between 1277 and 1279, Bacon was condemned to prison by his fellow Franciscans because of certain suspected novelties in his teaching. The condemnation was probably issued because of his bitter attacks on the theologians and scholars of his day, his excessive credulity in alchemy and astrology, and his penchant for millenarianism under the influence of the prophecies of Abbot Joachim of Fiora, a mystical philosopher of history' [8]. This is probably an over-simplification.

This time he had ink and vellum, and he wrote madly. Some of what he wrote was abuse of the Order and of contemporary theology and philosophy, but there was constructive writing too. He was called to a general chapter in Paris in 1278. He was condemned for novel and suspect writing and was prevented from writing to the Pope (known to be sympathetic). The behaviour of the Franciscan Order requires explanation. While they did not resort to torture or to physical violence, their policy of just locking up difficult cases like Bacon requires elucidation. It was so completely contrary to the spirit of St Francis that there must be something unexplained. It would seem that there was, in the religious orders in particular, a terror of the preternatural. Some Orders, noticeably the Carmelites, for some reason, escaped this, but it was very widespread. This was nothing to do with witchcraft; the fear of that did not become strong until shortly before the Reformation, and reached a peak in the seventeenth century.

2.2.2. Attraction in Bacon's thought

On first seeing Bacon's ideas about attraction the modern physicist is amazed. It seems that Bacon has discovered the exchange theory of nuclear forces, generally attributed to Yukawa in 1935. The particle theory of radiation, or quantum theory, is also there. But is it really? Well, it is in a way, in so far as radiation is explained as an emission of particles, just as in quantum theory. Moreover he had discovered the universal force of gravitation: Lindberg [9] remarks that 'the term has been appropriated by Grosseteste and Bacon to denote Al-kindi's universal force radiating from everything in the world to produce effects'.

Lindberg reminds us, however, that Bacon is a thirteenth-century man, no matter how modern some of his thought might appear. There are insights in Bacon's work which go far beyond the Greeks, and far beyond the thirteenth century. At a time when theology was becoming more complex, Bacon was pleading for simplicity. At a time when Scripture was becoming less important, Bacon was asserting the need for teaching simple Scripture both to clergy and laity. Thus he anticipated Luther by three hundred years. But it is easy, with such diverse material, to pick out passages which seem modern.

Apart from some excesses, his science was sophisticated. His insistence on experiment was new; though observation followed the Greek tradition. As we have said, his idea of experiment resembled chemistry rather than modern physics or astronomy. Nevertheless it was an organized systematic procedure, and it used mathematics, the hallmark of modern science, though here it was limited. Page after page of the *Multiplicatione Speciorum* and the *Opus Majus* are filled with geometry of the Euclidean type, when a few lines of trigonometry would suffice.

However, his physics is by no means trivial. His concept of the multiplication of species strongly resembles Huygens' idea of secondary wavelets, in the seventeenth century, which was the basis of the wave theory of light, and is still used in teaching. It may be said that his multiplication theory had no basis in experiment, but then neither did Huygens', four hundred years later. Altogether we cannot entirely agree with Lindberg, expert as he is. To the physicist, there is much that is new in Bacon's work, and not derived from the Greeks. His rules for the focusing of concave mirrors break new ground. His work on spherical aberration is also new. His invention of spectacles, though never fully worked out, and a very close approach to the principle of the telescope, was new and strange in the thirteenth century. One problem was in the grinding of lenses. For this reason work with mirrors figures more largely in his optical writing. Also we must admit that in his treatment of lenses, particularly the ocular lens, he is far from clear about its function (*Opus Majus*, fourth distinction). Mirrors are much easier to handle mathematically than lenses.

In spite of this mathematical deficiency, he was capable of handling conic sections, and showed that parallel rays would be focused at a single point on a paraboloid of rotation.

2.2.3. His texts and the problem of heresy

Why was Bacon condemned after Clement IV died, and indeed imprisoned for two years? The citation says 'for numerous novelties'. Were these theological, philosophical or scientific? It seems that they were just generally novel and do not fall into any single category. Basically, Bacon was dissatisfied with his sources as corrupt. He wanted to use the later Arab writers as secondary sources for Aristotle. Averroes raises his head again. Why this fear of Averroes? Bridges [10] says: 'The pantheistic tendencies discernible in Averroes and other Arabian thinkers had been diffused, as we have seen, by men like Amaury and David of Dinant. They were responsible, as some thought, for the disastrous anarchy which early in the century had devastated southern France'. It is no longer believed, so far as the author knows, that Albigensianism was connected particularly to the later Arab writers. A similar problem which Bacon fortunately missed by dying, was the confrontation between the realism of the Dominicans and the nominalism of Duns Scotus and other Franciscans. We say fortunate in no frivolous sense, for Bacon would have found it very difficult to side with those of his own order. Probably, he would have brushed the whole controversy aside.

So why was Bacon imprisoned? The answer must lie in the *Opus Majus*. Bridges divides the work into seven sections [11]:

1. the four general causes of human ignorance and error—undue regard to authority, custom, popular prejudice, and a false conceit of our own wisdom;
2. the close affinity between philosophy and theology;
3. the utility of the study of foreign languages;
4. the utility of mathematical science;
5. perspective or optics;
6. experimental science;
7. moral philosophy.

Probably, the answer does lie in his science. Theologically he was impeccable, apart from demanding simpler expositions for the layman, and this was not unusual. His demand for greater use of Arab sources was perhaps dangerous, but Albert the Great and Thomas Aquinas had done the same. So it must have been his gunpowder or his optics that gave rise to his problems. Of course in his optics he had relied very heavily on Arab sources, particularly Alhazen, so the problem is not simple. Arab sources are probably involved, whatever the details.

Returning to his optics, it seems to lack only mathematics and the wave picture to make it seventeenth century: 'But a species once it has been multiplied in the medium requires only the medium, and by itself, from its own active power, it can produce its like'. It seems an excellent picture of the secondary wavelet model [12].

With regard to the very high velocity of light: 'We see that the Sun and the Moon alter the same quantity of the medium with equal speed; and a candle and a fire illuminate the same dark place with equal swiftness' (at the same time he does not exclude a very high but finite velocity) [13].

One of the recurrent issues throughout the Middle Ages was the reform of the calendar, which had grown to be about ten days out of step. This was beginning to be significant in planting crops and other activities. It was realized that the Julian date was wrong by a small but cumulative amount. The problem was largely how to formulate a suitable expression that would not be merely arbitrary (as the astronomer's calendar is now, though to sixteen decimal places of accuracy). This was not solved until the sixteenth century, but Bacon made a number of suggestions. It was a highly political move, and Britain was to lag behind Europe for more than a hundred years before reform. John Dee, however, in a memorial addressed to Queen Elizabeth in 1582, pointed out the contribution which had been made by Roger Bacon.

An important interest of Bacon's was in philology and comparative languages. He fought for the study of Greek, Chaldean, Arabic and Hebrew, in the universities, as well as experimental science. He himself probably knew modern rather than classical Greek. His knowledge of Arabic was limited, as he says himself: 'I have touched Arabic but cannot write it'. Arabic was in fact very little known in the West. It is said that a letter to St Louis from the Sultan of Babylon could neither be read nor answered by anyone in the University of Paris.

His belief in astrology is sometimes brought forward as a point against him. This seems unfair since as late as the seventeenth century Kepler, one of the greatest astronomers, was a devout believer and practitioner of the art. Moreover, Bacon brings forward arguments in favour of forces acting between the planets and individuals on the Earth. Which of course they do, but they are so tiny and undifferentiated that it is difficult to see them being effective. Nevertheless there is scientific thinking there, and not just superstition.

Several authors make the point that Bacon's views of the methods of experimental science are far more accurate and indeed modern, than the views of his namesake Sir Francis Bacon, who is often credited with important advances in the scientific method. The latter, indeed, sometimes seems naive by comparison. So far as we know, they were not related.

Altogether, we can find no particular instance of the 'novelties' for which Bacon was imprisoned—his whole life was a novelty—but his scientific work seems the most likely candidate, hence his appearance in this book.

2.3. Bruno and a multiplicity of universes

Of all the characters described in this book, Bruno must stand out as the most persecuted. He was excommunicated successively by the Catholics, the Calvinists and the Lutherans. In Rome he was unjustly accused of murder. He was thrown out of Oxford after a singularly nasty encounter with Aristotelian professors. In several places, proceedings for heresy were prepared against him. Finally, of course, the Inquisition got him, first the relatively liberal Venetian one, then the far from liberal Roman one. After seven years deliberation he was condemned and burnt at the stake in the Campo di Fiori, where today a massive statue of him stands. He was more mobile than even a modern solid-state physicist, frequenting, and usually being expelled from, practically every university town in Europe.

Reading his work is exasperating. At first he seems startlingly modern, apparently anticipating important pieces of current cosmology. Then he relapses into what seems gibberish of a cabalistic kind. However, what runs through all his writing is a dedication to freedom of thought, rare enough in the sixteenth century. But he was not a pleasant man; violently anti-semitic, even by the standards of the time, a viciously obscene male chauvinist, self-centred and self-satisfied.

He made a living, such as it was, by teaching mnemonics, to customers who were usually dissatisfied, and by proof-reading. His philosophy was generally considered too dangerous to risk releasing him on a class.

He was born in 1548, at Nola near Naples. In 1562 he went to the University in Naples, to study the humanities and logic. He was attracted by Averroism, by no means dead in spite of the Fourth Lateran Council. This led him to the idea of an infinite, eternal universe and to a plurality of worlds. His plurality of worlds is one of stars rather than of universes, and is based on Lucretius rather than Copernicus. He is, in modern terms, describing a galaxy, but even that was an incredible step forward. Unfortunately he had no concrete evidence to support his opinion, although he had the courage to make an enormous jump in calculating the size of the universe and even discussed the concept of infinity in astronomy.

But we are getting a little ahead of ourselves. In 1565 he entered a Dominican monastery in Naples. As a student he was suspected of heresy, but was nevertheless ordained, and sent back to the monastery to complete his theological studies. He was reading Erasmus and other condemned authors and the suspicions grew. He left and made for Rome. There he got mixed up in a strange affair and was accused, quite unjustly, of murder. A second trial for heresy was threatened and in 1576 he had to flee from Rome. In the next fifteen years he lived in nineteen cities and matriculated in eleven universities. The longest time that he stayed in any one place was two years, in protestant Toulouse, where he got his Doctorate. From there he moved to Paris, to a congenial atmosphere; but he stayed less than two years. Then he spent an explosive few months in Oxford. In England he became a friend of Sir Philip Sidney, that noblest of Englishmen, and met the queen. He was a reincarnation of the mediaeval wandering scholar, but more intellectual. He returned to Paris, but things had changed, and he found it less comfortable, so he moved to College de Cambrai where he got a job, but that too failed, and he went to Germany.

First he tried the very prestigious Protestant University of Marburg, the earliest and most distinguished of the German protestant universities. He lasted there three days, and never taught a class. From Marburg he went to Wittenberg, where he settled for a while, in Luther's University, from August 1586. Here he was happy, and more important, tolerated. After two years, however, the Lutherans at Wittenberg were superseded by the Calvinists and in March 1588 Bruno had to leave. The issue was the Copernican theory, which had been condemned by the Calvinists. However, there were other more deep-seated problems: with Bruno, there always were. In his *Oratio Valedicta* on 8 March 1588 he said, after praising Luther: 'The day the Germans fully evaluated the power of their genius and applied it in higher studies, they would no longer be men but gods' [14]. From Wittenberg he moved to Prague, where he was unemployed, apart from teaching his mnemonics. His stay was brief and he moved on to the Lutheran Helmstedt University. Generally he was happier with the Lutherans than with the Calvinists, for while both Luther and Calvin had condemned Copernicus, Calvin was far more militant. The Catholics too regarded Copernicus as heretical, but for them the issue was not yet live because of their different emphasis on Scripture. Ten years would see a change.

In Helmstedt, Bruno matriculated on 13 January 1589 [15] and lived happily there for a year, in so far as this tortured man can be said to have lived happily anywhere. Unfortunately war broke out between Lutheran and Calvinist factions in the spring of 1590, and Bruno left for Frankfurt. This was at the time a very active centre of book publishing. Bruno had written innumerable tracts. The university senate decided solemnly that he could not lecture, or even stay. He eventually found refuge in a Carmelite monastery. In February 1591, Bruno, for some unknown reason, left for Zurich while his work *De Minimo* was being finalized. A friend finished it off and published it. After a brief absence he returned and prepared to publish *De Monade* and *De Immenso,* which with *De Minimo* are generally considered his most important works.

Following this, a Venetian nobleman, Giovanni Mocenigo, invited him to return to Venice, and he accepted. It was a crazy decision; he had already come to the attention of the Neapolitan and Roman Inquisitions. Nevertheless, he went. Was he seeking martyrdom? His whole behaviour had always been bizarre, and apparently regardless of personal safety.

He soon fell out with Mocenigo, who found his mnemonic teaching unsatisfactory. (It is perhaps necessary to explain why memory was so important; far more than it is today. There was a vast body of classical works, and commentaries upon them. Modern simplified techniques such as the calculus had not been invented. Memory, it was said, was power.)

The Chair in Mathematics in Padua was vacant, and Bruno was hoping to obtain it. There was in fact little chance. The electors had no intention, in spite of the rift with Rome, of appointing a heretic. Moreover they had another man in mind, Galileo Galilei, and he duly took up the Chair in 1592. We will deal with his fortunes later.

As for Bruno, Mocenigo betrayed him to the Inquisition with a plethora of heresies:

- The Catholic Faith is full of blasphemy against the majesty of God.
- There is no distinction of persons in God.
- The world is eternal (fourth Lateran and Averroes again).
- There are infinite worlds.
- The world is guided by fate.
- Souls pass from one animal to another.
- Christ was a rogue.
- Christ and his disciples were magicians.
- Catholic doctrines are asinine.
- Transubstantiation is a blasphemy.
- There is no punishment of sins.
- The virgin birth is impossible.

Whether Bruno really believed in all these things we do not know; Mocenigo was a dissatisfied customer, who seems by all accounts to have been a most unpleasant individual. However, any one of these heresies was enough to send poor Bruno to the stake.

The Inquisition struck swiftly. On 26 May 1592 his trial began in Venice: 'Hence, motivated by the sombre realisation that the heretical premises of his Weltanschauung made him vulnerable to the relentless assaults of his prosecutors, he alternately denied or justified or offered apologies for many audacious and original ideas ... in a desperate attempt to save himself, he was humbly to admit that he was in error, plaintively ask forgiveness for his "misdeeds", plead for a *modus vivendi* with the Church, and throw himself upon the mercy of the tribunal' [16]. He does not appear heroic; but which of us would?

In fact the Venetian Inquisition was relatively lenient. Out of 1065 people tried during the sixteenth century, only five had been sentenced to death. But Bruno had not reckoned with a new Pope. Clement VIII had promised strong measures against heretics, and on 12 September 1592 the Roman Inquisition demanded the extradition of Bruno. Venice delayed and procrastinated, but under great pressure, gave way; and in late February 1593 Bruno was transferred to the prison of the Roman Inquisition. At this point they seemed to lose interest, and, apart from occasional interrogations, did nothing for six years, until 1599. Then on 14 January he was presented with eight heretical doctrines taken from his works and invited to recant. Six years of imprisonment had stiffened his resolve. He refused to recant: 'He did not want to, nor did he wish to retract; nor was there anything to retract' [17]. If Mocenigo's allegations were true, there was quite a lot that the Inquisition would want him to retract. We should not, however, take Mocenigo too seriously. He was, as we have said, a dissatisfied customer. Moreover he seems to have borne some personal grudge against Bruno. If Bruno really had confided these beliefs and disbeliefs, he was suicidal, and the evidence is against this. There is, however, some evidence for mental unbalance in Bruno. Well authenticated statements that he made throughout his life, showing wild disagreements with both Catholic and Protestant teaching, manifest a complete disregard for his own safety; and yet from his final statements he was not looking for martyrdom. There is a total dissociation between what he said and what he had written. Perhaps he simply realized too late what he had done.

Bruno's astronomical speculations must not be confused with those of Tycho Brahe, Kepler or Galileo. He was not an astronomer. So far as is known he never made an observation. His support for Copernicus seems to have been based on philosophical or necromantic grounds. But his triumphant cry: 'there is an infinity of worlds, and an infinity of universes', haunts us with its modernity. The variety of offences for which the scoundrel Mocenigo denounced him were, in fact, probably largely true, at least in the philosophical field; though he claimed at his examination to be an orthodox Catholic theologically. At his trial he seems to have been hardly present in spirit; there is a desperate sadness about the event which affects us all. Yet at the end he shows that he was there: 'You perhaps pronounce sentence against me with a fear greater than that with which I receive it'.

On 16 February 1600 he was taken to the Campo di Fiori in Rome, gagged to prevent him making a speech, and burnt at the stake. It is said that he was offered an image of the crucified Christ, but rejected it. But this is hearsay. The anti-clerical Victor Emmanuel government, in the late nineteenth century, erected an enormous statue of Bruno where he was burnt in the Campo di Fiori. Its main object was presumably to annoy the Pope, but Bruno has not been forgotten. Always there is a wreath at the foot of the statue, and most days there are fresh flowers.

What is it about this strange figure that makes him so hypnotic? His thought is mediaeval but its tone is strangely modern. Above all it is free, free in the best sense of the word. Even those bound by allegiance to the Catholic or other faiths, can feel this freedom. He was condemned by all, now he belongs to all.

2.3.1. His philosophy and theology

He had a fantastic memory himself, and tried to systematize mnemonics, for teaching to clients. However, as we have seen, most of them seem to have been dissatisfied, which led to his downfall. Otherwise he made a precarious living by reading proofs, and occasionally, as in Toulouse, was appointed to a lectureship in philosophy at the university. His financial high-point was in Paris, where he was appointed *un lecteur Royal,* a considerable distinction. For some reason he gave this up to move to London. Apart from works on mnemonics and a play in Italian, he published nothing of importance during his stay in Paris. In London he began his dialogues in Italian. There are six; three of them are on the universe, and three are on morality. From our point of view perhaps the most important is the first: *Cena de Ceneri,* or *Ash Wednesday Supper* (Ash Wednesday, of course, is a day of fasting). In it he supports the Copernican theory, but goes further and suggests an infinite universe with an infinity of worlds each more or less like the solar system. The title is significant; a Lenten hardship is implied throughout. It is difficult for the modern reader accustomed to abstraction in scientific writing to appreciate the more discursive style of Bruno, and indeed, after him, Galileo. People intrude all the time, but without being obtrusive. In *Cena de Ceneri* the Ash Wednesday theme is always present. He makes important statements about the interpretation of Scripture, which anticipate Galileo by thirty years. He criticizes general behaviour in England and particularly in Oxford. Whatever the behaviour of the Oxford professors, those two years were the most productive in his life.

His second book written in England was more philosophical: *Concerning the Cause, Principle and One.* He explores the relationship between Matter and Form. To a modern non-specialist reader this book is probably rather obscure. We will not consider it. His third book, however, we cannot ignore: it is *De l'Infinito Universo e Mondi* (*On the Infinite Universe and the Worlds*) (1584). This is a disciplined and highly systematic critique of Aristotelian cosmology in which he develops his own model of an infinite universe. It is not clear whether he was fully aware of the work of Aristarchus, the greatest astronomer of antiquity, who was the only person to recognize fully the significance of the stellar parallaxes, or rather the absence of stellar parallaxes. If the Sun were really at the centre of the solar system, stars should show relative movement during the year. Nothing was observable at that time. Therefore, either the heliocentric theory was wrong, or the stars were a very great distance away. Copernicus was aware of this problem; it is not clear whether Bruno was. Bruno's reasoning was Aristotelian, though he rejected Aristotle. For instance in his *On the Infinite Universe and the Worlds* (1584) he writes [18] (in the following, E = Elpino, P = Philotheo, F = Fracastoro):

E: How is it possible that the universe be infinite?
P: How is it possible that the universe be finite?
E: How do you claim that you can demonstrate this infinitude?
P: Do you claim that you can demonstrate this finitude?

E: What is this spreading forth?
P: What is this limit? ... if the world is finite and if nothing lieth beyond, I ask you WHERE is the world? WHERE is the Universe? Aristotle replieth, it is in itself ... [But] position in space is no other than the surfaces and limit of the containing body, so that he who hath no containing body hath no position in space. What then dost thou mean, O Aristotle, by this phrase, that 'space is within itself'? What will be thy conclusion concerning that which is beyond the world? If thou sayest, there is nothing, then the heaven and the world will certainly not be anywhere.
F: The world will then be nowhere. Everything will be nowhere.
P: The world is something which is past finding out ... if thou sayest that beyond the world is a divine intellect, so that God doth become the position in space of all things ... I say [that it is] impossible that I can with any true meaning assert that there existeth such a surface, boundary or limit, beyond which is neither body, or empty space, even though God be there. For divinity hath not as aim to fill space, more therefore doth it by any means appertain to the nature of divinity that it should be the boundary of a body.

The argument is somewhat obscure, but it is original. If there is no edge to the universe, there can be no centre. Although Bruno's suggestion is original it is not the first suggestion of the infinite heliocentric universe. This came from Thomas Digges, the English astronomer, who was more scientific in his expression of ideas. His work was influential in England, interpreting Copernicus for the layman in the late sixteenth century, but Bruno does not seem to have been aware of it.

In the rest of this tract he develops an Averroistic view of the universe, of which the text quoted is part; but he also formulated another Averroistic view of religion, according to which religion is for the poor and un-lettered, whilst the educated may rely on philosophy for the purpose of government. Considering the conflict between the crowned heads of Europe and the Papacy at this time, it is difficult to imagine anything more explosive.

He was in great form that year. The next tract was *The Expulsion of the Triumphant Beast* (1584). This has been splendidly translated by Imerti [19]. It is historically important because it was the only one of his books to be cited by the Inquisition prosecutors; presumably because it contains all or most of his heresies. It is an allegorical work, in which at various degrees of obscurity, established institutions are attacked: the gods are dying; Jove in particular is decrepit, his teeth falling out, after an eternity of libidinous and other misdeeds, similarly, to a slightly lesser extent, for the other gods, and their misdeeds are immortalized in the constellations of the heavens. Jove calls a meeting at which he proposes to expel vice, the triumphant beast, from the sky, and substitute the virtues. It was a religious satire on superstition, but was also an attack on Calvinism, and particularly the principle of salvation by faith alone. But it goes further and is profoundly anti-Christian. Christ himself is Orion [20]:

> He knows how to perform miracles and can walk over the waves of the sea without sinking, without wetting his feet, and with this, consequently, will be able to perform many fine acts of kindness. Let us send him among men, and let us see to it that he give them to understand all that I want and like them to understand: that white is black, that the human intellect, through which they seem to see best, is blindness, and that that which according to reason seems excellent, good, and very good, is vile, criminal and extremely bad. I want them to understand that Nature is a whorish prostitute, that Natural Law is ribaldry, that Nature and Divinity cannot concur in one and the same good end, and that the justice of the one is not subordinate to the justice of the other, but that they [Nature and Divinity] are contraries...

In this passage Bruno seems to go beyond an attack on Christianity and even beyond the Manichean position.

Generally in this book he succeeds in insulting practically everyone. The Irish, for instance [21]: '"How are we to deal with this terrible Dragon, oh Jove?" asked Mars. "Let Momus speak," answered the father. "He is a useless beast and better dead than alive" responded Momus. "However if it seems to you that we should do so, let us send him to graze in Hibernia or one of the Orkneys. But be most careful [with the natives]".' Thus were the constellations re-arranged.

Other works at this time were *The Cabal of the Horse Pegasus* (1585) which dealt with the relationship between the human soul and the universal soul. It was a pantheist work and highly heretical in almost all denominations. Queen Elizabeth had said 'I do not seek a window which will look into men's souls'. It was fortunate that Bruno was living in England, for the heretic hunters would have persecuted him for this book, more than any other that he wrote. The *Heroic Frenzies* (1585) continues the pantheist strain, but is more high-minded and given to exhortation. From what would have seemed a safe haven in England, Bruno suddenly went back to France, but a France greatly changed. It is not clear why. England would have seemed the one country in western Europe where Bruno would have been safe. Perhaps he thought about it; perhaps he did not. Of course the excommunication pronounced by Pius V against Elizabeth did not help. All foreigners were suspect, even heretics against the Catholic faith.

In 1585 Bruno returned to Paris. Things had changed as we have seen, but he still went on writing. After some minor skirmishes Bruno launched a full-scale attack on the Aristotelians. The Catholic moderates thereupon rejected him and he had to leave Paris.

Thereupon he went to Germany. He still published work, but now it was minor and obscure. His great works: *De Monade*, *De Minimo*, and *De Innumerabilibus sive de Immenso* go back to the atomism of Democritus, and are important in the history of science because of that. These works were completed in Frankfurt, where the Carmelites gave him refuge, somewhat to the apprehension of the prior, who thought that Bruno did not possess 'a trace of religion'. Nevertheless they gave him refuge. After this work in Frankfurt in 1590, he did not publish again but in August 1591 made the fatal mistake of returning to Italy. England, France and Germany had given him refuge, but Italy, even anti-clerical Venice, was not ready for him. The torch would be picked up by Galileo, a greater man.

2.3.2. Postscript

It has very recently, and very plausibly, been argued that Bruno spent his time in England as a spy for the English against the French. There was, as usual in Elizabethan England, a plot against the queen. In this case it was a combination of the French and the English Catholics. Bruno, it is said, betrayed the plot to the authorities from the safety of the French Embassy, where he had been given generous hospitality. This cannot be proved, but it modifies our view of the man. Instead of manic and extravagant, he becomes cold and calculating. It is for the reader to decide.

It may be questioned whether Bruno belongs in this book. He made no observations, he made no calculations, but the scientific importance of the principle he put forward was immense. Rowan-Robinson, in his book *Cosmology*, says [22]:

> It is evident that in the post-Copernican era of human history, no well-informed and rational person can imagine that the Earth occupies some special place in the universe. We shall call this profound philosophical discovery the Copernican principle, although the first clear statement of it is due to Giordano Bruno. The discovery of millions of stars like the Sun, of other possible planetary systems, and of galaxies similar to our own Galaxy, all help to convince us of the truth of the Copernican principle.

Chapter 3

GALILEO AND THE INQUISITION

The Galileo controversy had three aspects: astronomical, philosophical and religious. These three strands were intertwined so that it is difficult to separate them. The astronomical argument was over the model of the universe put forward by Copernicus against the traditional model of Ptolemy. The philosophical argument was over the nature of celestial matter as laid down by Aristotle. The religious argument was over the interpretation of Scripture. Of these the first and the third are the most obvious; but a number of authorities hold that the basic conflict was between Galileo and entrenched Aristotelians in the universities.

3.1. The younger years

Galileo Galilei was born on 15 February 1564 in Pisa in the Italian duchy of Tuscany, to a family of noble ancestry, in somewhat reduced circumstances. His father was Vincenzio Galilei, a musician with a small income but a wide reputation as an innovator and authority on musical theory. In later life Galileo was to incorporate some of his father's discoveries on stringed instruments into his own work. Their relationship seems to have been good; better than that of Galileo with his mother. After a brief spell as a novice and then teaching in a monastery at Vallombrosa, he enrolled in medicine at the University of Pisa. He already knew that medicine would not be his life's work, that his talents lay in mathematics and the physical sciences, but it was his father's wish that he should follow a relatively safe career. He did not complete the course, and left without obtaining a degree, but made a reputation for teaching mathematics as a freelance. He was an abrasive character, capable of great loyalty, but could also be personally very offensive. He had a talent for writing funny but scurrilous verses about his colleagues and enemies. Pictures of him as a young man are rare; most show him in middle years, of medium height, heavily built,

Figure 3.1. *Nicholas Copernicus (1473–1543).*

bearded and pugnacious. He was especially forthright in his opposition to Aristotelianism, although he himself had been brought up on Aristotle.

Aristotle (384–327 BC) having been almost unknown in the West for more than a thousand years, enjoyed a revival in the twelfth and thirteenth centuries, due to the translation of his major works from Arabic into Latin. The Dominicans, Albert the Great and Thomas Aquinas, built a magnificent intellectual structure of scholastic philosophy and theology, based on Aristotle and Scripture, in the thirteenth century. This dominated Christian thinking for hundreds of years. Aristotle did not have a cosmological model of the solar system, except in the crudest terms, but he laid down the characteristics of celestial and terrestrial matter. Celestial matter was perfect, immutable and eternal. A celestial body must be perfectly spherical, without defect. Terrestrial matter, however, was degraded, subject to irregularities and decay. These categories were philosophical, though they had theological overtones. They had no basis in observational astronomy.

The model generally adopted by scholastic cosmologists was that of the Alexandrian astronomer Claudius Ptolemy (c90–168 AD). For its time it was a very remarkable scientific achievement. In this model the Earth was at the centre of the universe, motionless and without rotation (figure 3.2). The Sun, Moon and all the planets rotated about it. Beyond them were the stars. The whole universe rotated about the Earth every

Figure 3.2. *Harmonia Macrocosmica of Andreas Cellarius. Reproduced by permission of the Mary Evans Picture Library.*

day. The movement of the planets was not simple, however. To reconcile the model with observation the planets had to undergo a periodic retrograde motion, moving on epicycles (figure 3.3). In some versions of the model the planets were carried on crystal spheres, but this was rather fanciful and not essential to the basic idea. The Ptolemaic system was accurate to about half a degree, roughly the angular size of the Moon. In the early sixteenth century, Copernicus (Nicholas Kopernigk) showed that a simpler system was obtained by putting the Sun at the centre with the planets, including the Earth, rotating around it. The Earth revolved on its axis every day, removing the necessity for a diurnal rotation of the whole universe. Some epicycles were still necessary, but fewer than in the Ptolemaic system. Copernicus did not publish until just before he died in 1546, and the Lutheran clergyman Osiander inserted a preface, stating that the system was purely a mathematical model, without physical significance. This was not Copernicus' intention at all. Luther and Melancthon both dismissed the idea out of hand on the grounds of Scripture and of common sense. It made little impact until the end of the century, though Galileo was interested in it by 1595, and Bruno was campaigning for it well before this. The Danish astronomer Tycho Brahe proposed a compromise model in which the inner planets Mercury and Venus rotated about the Sun, which then rotated about the Earth with the outer planets.

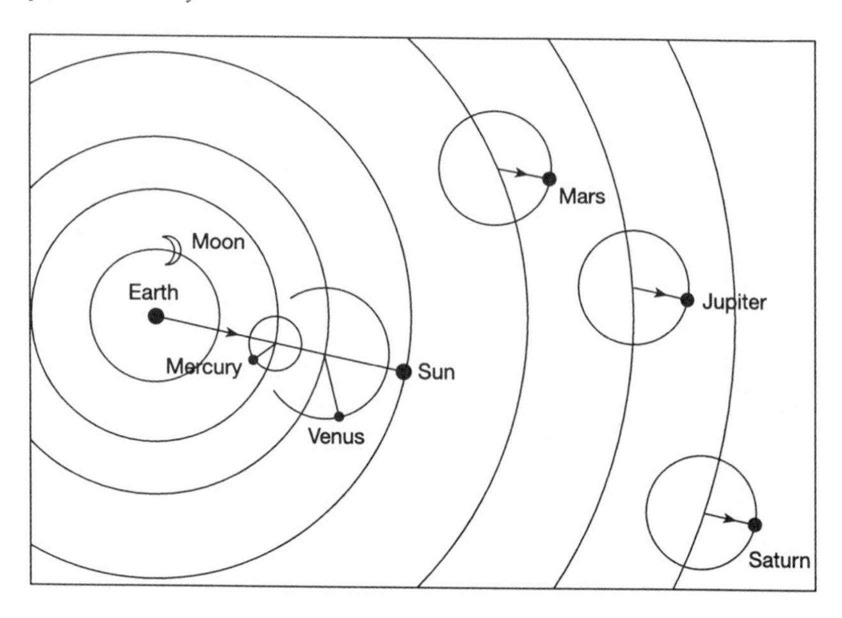

Figure 3.3. *Epicycles.*

At the age of twenty-five Galileo was appointed to the Chair of Mathematics in Pisa. It was very badly paid, but he stayed there for three years, recruiting the first of the ever-growing band of enemies that were to pursue him all his life. In 1592 he was appointed to the vacant Chair of Mathematics at Padua. Here he found a more congenial, anti-clerical atmosphere, and made many friends. Amongst them was Fra Paolo Sarpi, the theoretician of Venetian anti-papalism, and author of a history of the Council of Trent. This book was an essential component of the library of all Protestant scholars throughout the seventeenth, eighteenth and to a lesser extent the nineteenth centuries. A copy has been found bearing the signature Jonathan Swift, 1714. Sarpi was a very influential person in his time. Several assassination attempts were made by the Roman party. It is to the credit of Cardinal Robert Bellarmine, that he warned Sarpi of such attempts, whilst in the heart of the Roman party himself. They had been close friends before the split between Venice and the Papacy. It was Sarpi who was to advise Galileo not to return to Florence in 1610.

In Padua Galileo enjoyed himself and took a mistress, begetting two daughters and a son. He treated these badly. His common-law wife, Marina Gamba, was abandoned by Galileo in 1610 when he returned to Florence. She married in 1613. The son was legitimized, but the girls were put in a convent at Arcetri, a name which will recur, and were brought up under conditions of near-starvation. It was not a case of deliberate cruelty; the whole convent was half-starved. Livia, the younger, died in childhood. Virginia, the elder, survived to become a professed nun, and her father's greatest support in his old age. Her name in religion was Suor Marie Celeste.

Galileo was eventually appointed Chief Professor of Mathematics in 1596, at a salary of 180 florins a year. This was not much more than he had been earning in Pisa. He was thirty-two. Nowadays that would be considered almost middle-aged for a mathematician. Then, he was a prodigy.

Learning Greek mathematics, without the help of the calculus and other techniques, took a long time. Galileo mastered the cumbersome methods without difficulty. Recent studies indicate that Galileo was extremely active during these Padua years. Much of the experimental work which was afterwards used in the *Two New Sciences Dialogue*, was carried out in Padua. Indeed even a cursory look at the dialogues, published in 1638, suggests that this work was done by a young man, not a blind man in his seventies. In his later years in Padua his financial position became difficult because of the demands of his brothers and sisters. He agreed together with his brother Michelangelo, to pay the dowry for his sister Livia. It was an enormous sum, more than double his salary. Michelangelo never paid his share, indeed borrowed money from Galileo, moved to Poland, and eventually sent his whole family to live with Galileo in Padua. Galileo had to borrow heavily and secure an advance on his salary. In spite of these worries he was working hard and fruitfully.

Antonio Favaro, the great Galileo scholar who prepared the national edition of Galileo's works, completed in 1908, says in his introduction to the first American edition of the *Two New Sciences Dialogue* [1]:

> One who wishes to trace the history of this remarkable work will find that the great philosopher laid its foundations during the eighteen best years of his life—those which he spent at Padua. As we learn from his last scholar, Viviani, the numerous results at which Galileo had arrived while in this city awakened intense admiration in the friends who had witnessed various experiments by means of which he was accustomed to investigate interesting questions in Physics. Fra Paolo Sarpi exclaimed: to give us the Science of Motion, God and Nature have joined hands and created the intellect of Galileo. And when the *New Sciences* came from the press one of his foremost pupils, Paolo Aproino, wrote that the volume contained much which he had already heard 'from his own lips' during student days at Padua.

Wallace, in a recent book [2], has gone further and investigated Galileo's sources for his early work. He finds, not surprisingly, that they are Aristotelian, but not completely so. He finds that they are largely scholastic, using the techniques and sources developed primarily by Aquinas and Albertus Magnus. Moreover, he concludes that they owe a great deal to the work of the Collegio Romano, the Jesuit University with which Galileo was so often to come into conflict, and the home of the unfortunate Father Grassi, whom we shall meet later.

Padua, under the relatively benevolent rule of Venice, was a haven for Galileo. Tolerant, anti-clerical, but religious, it fitted his personality well. One can only speculate on what would have happened if he had stayed there. He invented an analogue computer for military use in 1597, and lawsuits over it occupied him greatly. Otherwise his published scientific output was small until 1610, when he produced a pamphlet which caused a sensation, the *Sidereus Nuncius*, or *Starry Messenger*. It was reprinted and distributed all over Europe. He was then forty-five. It is interesting and strange that he apparently did not take seriously the great supernova of 1604, except to comment that it showed a change in the celestial sphere. Kepler collected luminosity observations from all over Europe on this object, enabling a light curve as good as any modern observations, but Galileo was not one of the contributors. This was not a consequence of their one-sided antagonism, which only developed later. It seems that Galileo was not very interested in astronomy at this time. He was soon to develop an interest, however.

In Holland, Lipperhey and others had invented the telescope. Galileo turned one on the stars. The quality of the images was atrocious (some of Galileo's telescopes still exist) but he saw things never before seen: the craters of the Moon, the moons of Jupiter, the phases of Venus. The last of these was the most important. The craters of the Moon, by their variable illumination, demonstrated mutability and a non-spherical surface in the heavens. The moons of Jupiter showed a miniature solar system, but proved nothing. But the gibbous phase of Venus was only possible in a heliocentric system in which Venus at least must rotate about the Sun. The system of Tycho Brahe, in which the inner planets rotate about the Sun, and together with the Sun rotate about the Earth, fulfils these requirements, as does the Copernican system. Typically, Galileo did not publish this openly but in a Latin anagram. Russell, Dugan and Stewart [3] say:

> According to the theory of Ptolemy, Venus could never show us more than half its illuminated surface, since, according to his hypothesis, the planet was always between us and the supposed orbit of the Sun. Accordingly, when in 1610 Galileo discovered that Venus exhibited the gibbous phase as well as the crescent, it was a strong argument for the Copernican theory. Galileo announced his discovery in a curious way, by publishing the anagram—
>
> *Haec immatura a me iam frustra leguntur*: o.y.

Figure 3.4. *Saint Robert Bellarmine (1542–1621). Reproduced by permission of the Mary Evans Picture Library.*

Some months later he published a solution, which is found by merely transposing the letters of the anagram and reads

Cynthia figuras aemulatur Mater Amorum

meaning 'The Mother of the Loves [Venus] imitates the phases of Cynthia [the Moon]'.

However, to understand this required some knowledge of solar system kinematics and optics. The other thing that was discovered by the new instrument was the multitude of stars outside the solar system. Whilst reminiscent of Bruno and his heresies, this did not attract the attention of the Inquisition. Looking through one of Galileo's telescopes, which the author has done, it is difficult to understand how they could see these things, but see them they did.

Cardinal Robert Bellarmine, made a saint and a Doctor of the Church in the twentieth century, looked through a telescope and was intrigued by what he saw. He asked for a report from Father Clavius and his colleagues, who confirmed Galileo's findings. Clavius was the senior astronomer in Rome. He could perhaps be regarded as a Copernican without the courage of his convictions, though that is not quite fair; at least he was open-minded. Unfortunately he had died before the real battle commenced. Galileo's pamphlet was an instant success, but again, cost him friends. As we have said, the *Starry Messenger* was a best seller,

Figure 3.5. *Galileo Galilei (1564–1642).*

but was also a fierce attack on Aristotelianism, as well as potentially a strong defence of Copernicus. It was at this point, perhaps, that we can say that the main conflict began, though it was not until 1615, with the Dominican Caccini's attack, that it came out into full daylight.

3.2. The middle years

In 1609 the Grand Duke Cosimo II, who had been a pupil of Galileo, invited him to come back to Tuscany and become court mathematician and philosopher in Florence. It was an attractive offer: five times his salary in Padua and very little teaching. Galileo accepted, partly because he was homesick for Tuscany. His Venetian friends warned him not to take it. Florence was too closely allied with the Papacy. Nevertheless he did. His first activity was a book on hydrostatics, published in 1612, which was an attack on the Aristotelian professors, and not calculated to make friends, though it certainly influenced people. It revived Archimedes' ideas and vindicated them.

The main enmity, however, centred on the heliocentric theory. A publication on the sunspots, which Galileo did not discover, though he thought he had, raised controversy because again it raised the possibility of mutability in the heavens. In this, as in hydrostatics, Galileo was supported by the cheerful Benedetto Castelli, a Benedictine who was professor at Pisa. Unfortunately the Benedictines were fighting with the Jesuits at the time, which did not make for good relations between Galileo and his Jesuit colleagues. But the most vocal opponent was the Dominican, Tommaso Caccini. In a series of violent sermons, Fra Tommaso attacked mathematicians in general and Galileo in particular. He is not entirely without interest to us. On Ascension Day 1615 he preached on the text 'Viri Galilaei...', which could be translated as 'You men of Galilee...', or alternatively 'Ye men of Galileo, why stand ye looking up into Heaven?'. De Santillana has remarked on the interest which the characters in the Galileo story have, by comparison with modern counterparts. Caccini was generally regarded as a fool, but he was a witty fool.

Galileo, now with almost diplomatic status, complained to the Dominican Master-General, who apologized, saying that he could not control thirty thousand Brothers of the Order. The favour of the Grand Duke, a staunch ally of the Papacy, made Galileo almost untouchable at this time. Even the shadow of Giordano Bruno, it seemed, could not affect him. But Rome was moving. Caccini had been to the Inquisition [4] against his Master-General's wishes. Galileo was ill, but decided to travel to Rome to defend his position. He was now fifty. What exactly was he facing?

3.3. The Inquisition

There had been sporadic persecution, mainly of Jews and Moslems, since the mid-eleventh century, and some were burnt, but this was largely mob rule spurred on by beliefs of ritual murder of Christian babies, beliefs which survive in Eastern Europe to this day. The heretical sect of the Cathari began to spread, and these too were persecuted, though again mainly through mob violence, rather than formal ecclesiastical authority. In 1231, Pope Gregory IX instituted the Inquisition to formalize and maintain Church control of proceedings. Initially the job was entrusted to the Cistercians; but they proved both incompetent and temperamentally unsuited, so the organization was taken over by the Dominicans. Innocent IV in 1252 reorganized the system, and for the first time allowed torture of suspects to extract confessions, a thing which had been forbidden by a predecessor, who was afterwards canonized. The Inquisition was active in the battle against the Albigensians in southern France, in the middle of the thirteenth century. In the later Middle Ages it was virtually non-existent, except in Spain where it was reconstituted in 1478.

In the early sixteenth century, however, came Luther and the Reformation. Pope Paul III in 1542 re-established the Inquisition on a much more formal and legal basis. Its procedures were strict and canonical. According to the *Encyclopaedia Britannica,* its jurisdiction extended to all Catholic countries and was superior to bishops and princes. In practice, as in the case of Giordano Bruno, its effectiveness and severity varied widely from place to place. In the early seventeenth century it was very powerful in Rome itself. This was the organization that Galileo was to meet in 1615 and again in 1633. Redondi says [5]:

> Since 1569, the sober edifice alongside St Peter's had taken its name from the Supreme Tribunal and its adjacent prison on the ground floor of the Holy Office of the Congregation of the Supreme and Universal Inquisition. Founded in 1542 to centralize and guide the peripheral activities of the declining Inquisition, the Congregation of the Holy Office was the oldest and most important congregation of the Curia, so much so that the Pope kept its presidency for himself.... In charge of the assignment of imprimatur, and the palace's true 'grey eminence', however, was the Master of the Sacred Palace, a post traditionally held by a member of the Dominican order, who usually attained this post after having worked as a consultant.

This then was the complex and sophisticated organization that Galileo encountered. Over all the years in which the Inquisition operated it probably connived at the killing of about 200 000 people (mostly Spanish Christianized Muslims, who, having been dumped on the Barbary coast, were slaughtered by the Moors for their Christianity [6]), slightly less than Hiroshima and Nagasaki; considerably less than Auschwitz. So the twentieth century still holds the record.

(All these numbers have been disputed. In the case of the Inquisition, estimates of the numbers actually burnt in Spain vary from 200 to 30 000 with a majority view in favour of about 2000. The crimes against the Moriscos were not entirely due to the Inquisition—there was a great deal of popular support—but the Inquisition must bear a good deal of the blame.)

In fact the Inquisitioners did not carry out the executions, but handed the victim over to the secular arm with a ritual plea for mercy which was always ignored. As can be seen with Bruno, Venice was particularly lax in the early seventeenth century; Rome, under the influence of Bellarmine, extremely strict. All regional Inquisitions at this time had to employ readers, lay or clerical, to cope with the flood of books which had erupted on to the market with the rise of printing, and were looking for the imprimatur. Technically, all books, not just theological ones, were required to obtain an imprimatur. While this was sometimes waived, supervision of the process was strict. For a person of the standing of Galileo, a major official would examine his book, relying on theological assessors if there were any question of suspected heresy. The Protestant churches had corresponding, though much less elaborate, machinery. A major aim of the Inquisition was to centralize control of religion, in so far as possible, in Rome, and to take it away from local bishops and princes. This certainly made the processes of religious trials less arbitrary, and perhaps, strangely enough, on the whole more liberal, with the exception of the Spanish Inquisition. It is difficult for the modern reader to visualize how political religion was in the sixteenth and seventeenth centuries. While there was some crossing of boundaries, particularly in the struggle between France, Spain and England, generally the religion of the prince became the religion of his or her subjects; and this was enforced. In Protestant countries of course, the Inquisition was not active, but the enforcement was no less effective. The attempt of early Elizabethan England to pursue a moderate path was sabotaged from within and without, and finally disappeared after the papal excommunication of Queen Elizabeth.

This is not to say that the religion of the prince, if Catholic, automatically ensured an allegiance to the political policy of the Papacy. Venice, Spain and France all deviated, depending on who was occupying the Chair of Peter at the time.

De Santillana makes the point that Galileo, like very many of his compatriots, was Catholic, devout, but anti-clerical, and felt no problem in opposing or evading papal policies if he saw fit [7]. Until recently, this was a rare type in northern Europe, but seems to be on the increase, though perhaps not fast enough to counteract the flight from all forms of religion.

3.4. Galileo in Rome

In Rome, he was treated almost as royalty, though the Florentine ambassador greeted him with trepidation. The latter, Guicciardini, would nowadays be diagnosed as a case of anxiety neurosis; though in his position anyone would be anxious. Galileo had friends in Rome, among them the influential Cardinal Maffeo Barberini, who was to become Pope. He also had many enemies, whose number he increased by his merry jibes and witty remarks. In some respects this great man was entirely lacking in common sense, in others his perceptions were acute. In the *Letter to the Grand Duchess Christina* [8] he laid down principles of scriptural scholarship that were not accepted by the Roman Catholic Church until the encyclical *Providentissimus Deus* of Leo XIII in 1893, and by most Protestant Churches at about the same time, although some resist them still.

On 24 February 1616, the consultants to the Holy Office, having been requested by the Pope to judge on the Copernican theory, pronounced their verdict. They said unanimously that it was foolish and absurd philosophically and formally heretical as contradicting the Holy Scriptures. On the proposition that the Earth moves and rotates, it was foolish and absurd philosophically. This was the first of five interconnected documents associated with the 1616 discussions. Galileo was ordered by the Pope to meet with Cardinal Bellarmine privately on

26 February. Bellarmine had already discussed the situation with others. Talking to Monsignor Piero Dini, a friend of Galileo's, he said:

> He did not think the work of Copernicus would be prohibited, at most some saving addition might be made to the effect that it was only meant to explain appearances, or some such phrase, and with this reservation Signor Galileo would be able to discuss the subject without further impediment...

Also he wrote to Antonio Foscarini, provincial of the Carmelites in Naples (he had written a pamphlet on Copernicus) [9]:

> It seems to me that Your Reverence and Signor Galileo act prudently when you content yourselves with speaking hypothetically and not absolutely, as I have always understood Copernicus spoke.

So a hypothetical treatment would be acceptable. Cardinal Maffeo Barberini, the future Pope Urban VIII, spoke similarly.

With regard to the meeting on 26 February, in the Galileo file (opened to scholars as late as 1870), there is a strange note, in minute form, unlike the other documents in the file, and somewhat scrappy in appearance. It is dated 26 February, and purports to be a report on what happened that day. Even without the other documents which we will describe, this one is anomalous. It is worth quoting in full since it is the basis of the condemnation eighteen years later (Brodrick's translation) [10]:

> At the palace, the usual residence of the afore-mentioned Lord Cardinal Bellarmine, the said Galileo having been summoned and standing before his Lordship, was, in the presence of the Very Reverend Father Michael Angelo Seghizi de Lauda, of the Order of Preachers, Commissary General of the Holy Office, admonished by the Cardinal of the error of the aforesaid opinion and that he should abandon it; and immediately thereafter, in presence of myself, other witnesses, and the Lord Cardinal, who was still in the room, the said Commissary did enjoin upon the said Galileo, there present, and did order him, in the name of His Holiness the Pope, and the names of all the cardinals of the Congregation of the Holy Office, to relinquish altogether the opinion in question, namely that the Sun is the centre of the universe and immovable and that the Earth moves; nor henceforth to hold, teach, or defend it in any way, either orally or in writing. Otherwise proceedings would be taken against him in the Holy Office. The said Galileo acquiesced in this ruling and promised to obey it.
>
> Done at Rome, in the place aforementioned, in the presence of the Reverend Badino Nores from Nicosia in the Kingdom of Cyprus, and Augustino Mongardo, of the diocese of Montepulciano, both witnesses belonging to the said Lord Cardinal's household.

There are several peculiar things in this document apart from those which we will discuss later in the appendix at the end of this chapter:

(a) It does not start at the beginning, 'the afore-mentioned Lord Cardinal...'
(b) It looks like a rough draft for a formal minute, yet it is witnessed.
(c) The witnesses have been identified. They were servants in Bellarmine's house—no more. Yet according to the documents there were high-ranking officers of the Inquisition present. Why did they not witness the document?
(d) With all its faults the Inquisition was extremely scrupulous about its paperwork. This document could never have been produced in court, and later Inquisitors knew it. It purported to be an abjuration but was not notarized, or signed by Galileo.

What does the document say? Galileo must 'relinquish altogether ... nor henceforth to hold, teach, or defend it [his opinion] in any way, either orally or in writing'. It is very comprehensive. But is it? There is a whole section missing at the top: 'Admonished by the Cardinal of the error of the afore-said opinion'. What opinion? What it does say clearly, however, is that Galileo must not hold, teach, or defend in any way, the Copernican doctrine. This presumably forbids the suggestion of Maffeo Barberini that discussion *ex hypothesi* would be acceptable as a mathematical device. So he must give it up altogether.

Now this was not Galileo's impression at all. True, the Inquisition's consultants had made a comprehensive condemnation, but they had not ruled out holding Copernicanism as an hypothesis, and Galileo would settle for this, for the time being. So Galileo's recollection of his interview with Bellarmine differed radically from the secret minute. He stressed the courtesy with which he had been received, the civilized conversation, the discussion of possibilities. True, Copernicanism had been condemned, but in such a way that left openings. Yet the official documents, whose authenticity has never been questioned, are strongly hostile. The Holy Office met on 3 March. They said [11]:

> The Lord Cardinal Bellarmine having reported that Galileo Galilei, the mathematician, had according to instructions of the Sacred Congregation, been admonished to abandon the opinion he has hitherto held, to the effect that the Sun is the centre of the spheres and immovable, and that the Earth moves and had acquiesced therein; and the decree of the Congregation having been registered, by which were suspended and prohibited respectively the writings of Nicholas Copernicus *De Revolutionibus Orbium Coelestium*, of Diego Di Zuniga on the Book of Job and of Paolo Antonio Foscarini, Carmelite Friar—His Holiness ordered this edict of suspension and prohibition respectively, to be published by the master of the sacred palace.

The arrogance is monstrous, but there is no mention of teaching the idea as an hypothesis. Hence Galileo regarded it as a victory. He had an audience with the Pope, unusually benevolent for Paul V, and prepared to return to Florence in a cheerful mood.

Rumours were flying, however, that Bellarmine had admonished him to give up Copernicus completely. This is indeed the way that the edict of the Holy Office could be read. So Galileo appealed to Bellarmine, and received a certificate [12]:

> We, Robert, Cardinal Bellarmine, having heard that Signor Galileo Galilei has been calumniously reported to have abjured in our hand, and moreover to have been punished with a salutary penance, and having been asked to make known the truth as to this, declare that the said Signor Galileo has not abjured in our hand, nor in the hand of anyone else here in Rome, nor so far as we are aware in any place whatever, any opinion or doctrine held by him: neither has any penance, salutary or otherwise, been imposed on him. All that happened was this. The declaration made by the Holy Father and published by the Sacred Congregation of the Index was intimated to him, wherein it is declared that the doctrine attributed to Copernicus that the Earth moves round the Sun and that the Sun is in the centre of the universe and does not move from east to west, is contrary to the holy scriptures, and therefore cannot be defended or held.

These documents are confusing, for they all seem to be saying more or less the same thing. There are differences, however, which were important to Galileo, and indeed to others. The letter from Bellarmine absolves Galileo from any blame, and denies that he was forced in any way to recant. The secret minute, however, not only says that he was forced to recant but puts a total prohibition on all discussion of Copernicanism. The consultant's report condemns the doctrine as a physical reality, but does not mention anyone except Copernicus and does not expressly recommend a prohibition of discussing it as a mathematical hypothesis. The Holy Office condemns a number of works, but none of Galileo's, not even his book on the sunspots. To Galileo, who did not know of the secret minute, these were vital differences. He took them to mean that he was free to discuss the hypothesis *per se*, providing that he did not ascribe physical reality to it. It is interesting that the condemnations of 1616 concentrated on Copernicus and on the literal interpretation of Scripture. They ignored the anti-Aristotelian works such as that on the sunspots. Moreover, no work of Galileo was condemned, where one might think that *Sidereus Nuncius* (the *Starry Messenger*), was a prime candidate. Florence and the Grand Duke still carried more clout than the Universities, and Aristotle had already lost.

3.5. The later years

On returning from Rome Galileo became ill, and periodically was ill for the rest of his life. He slowed down; the books which had poured from him now took years to complete. He seems to have been working on different books at the same time, and began his long studies of the tides, which were to dominate much of the rest of his life. Galileo abominated action at a distance, which was built into Kepler's theory. Kepler was correct, attributing the tides to a combination of Sun and Moon, though a complete theory had to wait until the nineteenth century. Galileo preferred a more mechanistic theory, relating the orbital and diurnal motions of the Earth. This was a *cul de sac*, and led to much trouble later, since he was to rely heavily on it in the *Two Worlds Dialogue*. It was a tragic waste of time. Had he been more generous to Kepler it need not have happened.

In 1618 three comets appeared. Galileo was ill but his friends reported to him. In the Aristotelian world, comets were sublunary, but Tycho Brahe had demonstrated years before that a comet lay beyond the Moon. The Jesuit, Orazio Grassi, published a pamphlet, proving the comets to be a powerful argument against Copernicus. Galileo was enraged. According to Brodrick, himself a Jesuit [13]:

> The sad fact is that the Jesuit scientists were under orders from their General, Claudio Acquaviva, to stand by Aristotle, though their best men such as Clavius and Grienberger certainly leaned towards Copernicanism in their hearts.

Remember that this was the year of completion for Kepler. Those who had any detailed knowledge of astronomy should have known that Copernicus was right, and had been proved quantitatively right, with the transition from circles to ellipses. Galileo published the *Discourse on the Comets* in 1619. Maffeo Barberini, soon to be Pope, wrote in praise of it. Grassi replied with the *Astronomical and Philosophical Balance*, written under a pseudonym. For three years Galileo was silent. Then he replied in 1623, with a work called *Il Saggiatore* (*The Assayer*). Stillman Drake has called this 'The greatest polemic ever written in Physical science'. The book was published in 1623, the year that Barberini was elected to the Papacy, taking the name Urban VIII. Galileo dedicated his book to him. It was received very favourably; no one was going to criticize the favourite of a new Pope. Unfortunately, Galileo was quite wrong on the issue of comets, and Father Grassi was, almost, but not quite, right. Tycho Brahe had shown that the orbits of the comets could not be circular, and Galileo took this as an attack on Copernicus. Therefore the comets could not be real, but rather optical illusions caused by exhalations from the Earth. Kepler could have told him, had he chosen to listen, that no orbits about the Sun are circular.

The argument with Grassi, who published under the pseudonym of Sarsi, soon went far beyond that of comets. Urban VIII was delighted. He disliked the Jesuits and their Spanish leanings, and the sheer satirical brilliance of *Il Saggiatore* gave it a ready market in Rome. It was a best-seller. The distinguished Italian scholar Pietro Redondi, however, argues that it was Galileo's most heretical, perhaps his only, heretical book. It was in effect an atomistic tract. Atomism was the great controversy of Western philosophy, and sometimes had strong theological overtones. Since Aquinas had used Aristotelian physics to explore the mystery of transubstantiation, atomism was suspect. The sensitivity of the issue was heightened by Reformation speculation and controversy, in which, then, the theology of the Eucharist was paramount, before even the supremacy of the Pope. Curiously, no one in authority spotted this at the time, and Galileo's book was never cited before the Inquisition, except by an anonymous complainant [14].

It did, however, make many enemies beside the Jesuits. The complex, swirling world of Rome in which theology, philosophy, science and politics mingled and fought, produced enemies easily. Galileo now found himself, whether deliberately or not, arraigned against the whole Imperial, pro-Spanish faction. He probably did realize it, for he suddenly went quiet, though at the height of secular and papal popularity.

As with Gerard Manley Hopkins, there followed a long silence, of seven years. Galileo's equivalent of *The Wreck of the Deutschland* was his most influential, though not his greatest work: *The Two World Dialogues*. It was a powerful piece of polemic, written in the form of dialogue between Salviati, Sagredo and Simplicio: a Copernican, a moderator and an Aristotelian. He was working at it all the time, until 1630, but also suffered much from ill-health during the period 1620–1633. He had a hernia which grew progressively worse, and also had heart palpitations. In 1634 he would be seventy. (According to church reckoning he was seventy in his seventieth year, i.e. from 15 February 1633.) In terms of the life expectancy of the time, of course, he was lucky to have survived so long. Doubtless many wished he had not.

The *Dialogue* was finished in December 1629, but not published until February 1632. Its publication had not been easy for it required an imprimatur. Fortunately the Master of the Sacred Palace, chief licenser, was a Florentine and an old friend. Padre Niccolo Riccardi, an easy-going but anxious, enormous man, popularly known as 'The Monster', was concerned about the book. De Santillana says [15]:

> After that, it was time for Father Riccardi to go to work. The bustling 'Padre Mostro' went hastily over the manuscript and was not entirely reassured. He did not know much about astronomy, but the stuff did not look to him as hypothetical as he had been told. He delegated his assistant Father Raffaello Visconti to examine it and make such alterations as were needed. Father

Visconti, who was supposed to be versed in mathematics, went over the text, changed a few words here and there, and gave his approval. He had obviously understood insufficiently either the text or the papal instructions. Now the imprimatur from Rome was as good as granted.

But the imprimatur was not granted immediately. Father Riccardi was still not happy. Threatening noises were being made by the Jesuits and others. He could see trouble impending. He decided to read the manuscript himself a page at a time and pass it on piecemeal to the printer. But the printer could not start to print without a licence. Therefore Riccardi issued a licence before he had read the text. This was to cause endless trouble later, for an imprimatur is precisely a licence to print. Meanwhile he insisted that the Preface and conclusions should be re-written. Galileo left Rome in June, 1630 for fear of the heat, the malaria, and his enemies. He promised to return from Florence in the autumn with a new preface and conclusions.

Part of the work was examined in Florence and the new preface and conclusions were supplied. Riccardi was still unhappy. What happened next is well described by De Santillana [16]:

> Galileo had to rely now on the good offices of the Florentine ambassador. Luckily it was no longer Guiccardini who had the post but Francesco Niccolini, a faithful friend. He and his wife (she was Riccardi's cousin) had the 'Father Monster', as they called him affectionately, as a habitual guest, and now they bent their efforts on extracting a permission from him. Riccardi at first refused; then yielding to the subtle pressure of Caterina Niccolini, he relented and sent permission to have the final revision done in Florence. The Preface and the conclusion, however, he retained in his hands, 'to arrange them according to the wishes of His Holiness' [17, 18].

A somewhat different reaction came, predictably, from Florence [19]:

> The Florentine Inquisitor, Father Giacinto Stefani, read the book, changed a few words, but found nothing wrong with it. Indeed he was 'moved to tears by the humility and reverent obedience displayed by the author'. Another revision by the Inquisitor himself, Father Clemente Egidii, and still another imprimatur for the text; but a year had gone by, and still the preface and concluding statement were lacking. Galileo was in despair...
>
> Riccardi obviously did not dare come before the Pope again with his problem and ask for help concerning his wretched revision. He hung himself like an albatross about [Cardinal] Ciampoli's neck and asked for a direct order. He got his clearance. Even so,

he tarried. It was only on July 19 1631 'dragged by the hair', as Niccolini puts it, that he surrendered the packet to the embassy.

Thus, before publication, this book which was to be condemned underwent no fewer than three papal clearances for the imprimatur: Riccardi's in Rome, Stefani and Egidii's in Florence.

Anyway, it was done, and in February 1632 Galileo presented the first copy to the Grand Duke. It was a best seller though quarantine regulations over the plague which was still lingering, hindered its circulation. Two further imprimaturs were acquired, making five in all [20].

It is a powerful piece of work, which uses the strongest arguments but also the weakest ones. Galileo was obsessed with the tides, which are interesting but prove nothing. The gibbous phase of Venus and other strong arguments are hardly used. The choice of characters was unfortunate. Someone persuaded Urban VIII that Simplicio the Aristotelian was a caricature of himself (he had suggested one of the arguments which Galileo puts into Simplicio's mouth). Despite the unlikelihood of this, Barberini was enraged and became a violent enemy. He was, himself, in difficulties. The Spanish ambassador had accused him in open court of supporting the French and the Protestant Gustavus Adolphus against the Emperor. The Ambassador had Diplomatic Immunity, and the charge was perfectly true, so there was little he could do. He was looking for a diversion; perhaps this could be Galileo, but it seems that at this stage he was still prepared to give Galileo some support [21].

On 11 September 1632 the bombshell burst. Riccardi arrived at the Florentine Embassy in a panic and spoke to Niccolini:

> Under the seal of absolute confidence: there had been found in the books of the Holy Office that sixteen years ago Galileo had been summoned to Rome [incorrect; he insisted on coming] and forbidden by Cardinal Bellarmine in the name of the Pope and the Holy Office to discuss this opinion, and that this alone was enough to ruin him.

So no one knew of the existence of the document until 1632, even after two years of examination for the imprimaturs. One wonders did the document exist before 1632? (See the appendix at the end of this chapter.)

Urban VIII ordered that Galileo should be summoned to Rome for questioning by the Inquisition. The order came on 1 October 1632, so he was faced with a difficult winter journey. He pleaded that he should be questioned in Florence. This was refused. He appealed to Grand Duke Ferdinand, but the latter was terrified of Rome, though he did put his own litter at Galileo's disposal. Galileo obtained a medical certificate that his life would be put in danger if he travelled, but to no avail. All he could get was a postponement to the end of December.

Eventually he reached Rome on 13 February 1633. After a stay at the Florentine Embassy with his faithful friends the Niccolinis, he surrendered to the Inquisition early in April. He was officially seventy, and according to Inquisition rules, non-torturable. He was treated with respect, and given comfortable apartments instead of the common prison. He even had a servant, supplied by Niccolini. He was enfeebled, but remained confident. He had Bellarmine's certificate of 1616, which he believed gave him freedom to discuss Copernicanism as an hypothesis, but there were rumours of a second, more comprehensive document. This he did not understand. He was lonely. Benedetto Castelli and his other friends had been sent away from Rome, but Niccolini and his wife did what they could.

On 12 April 1633 Galileo was interrogated for the first time by the Commissary of the Inquisition, a Dominican named Fra Vincenzo Maculano da Firenzuola. It was a poker game. Galileo was intimidated by the suggestion of a new document but held on to Bellarmine's certificate. Firenzuola, on the other hand, knew that the document he had would not stand up in court—the Inquisition had quite strict rules of evidence. He had to extract some further confession from Galileo. So he returned to the interview with Bellarmine, and revealed his ace: 'he must neither hold, defend, nor teach that opinion in any way whatever'. Galileo denied ever hearing the last phrase; but his spirit was flagging, he was confused. At this point he collapsed and had to go to bed: remember he was nearly seventy years old and seriously ill. They left him for two weeks and then on 28 April Firenzuola reported success [22]—Galileo had recanted. He said that he had re-read the book and realized things in it which he had not seen before. He even offered to add additional chapters to disprove the Copernican hypothesis.

What makes this whole episode so appalling is that this was all fifteen years post-Kepler. Anyone who had studied the subject closely knew that Copernicus was basically right, though with ellipses instead of circles. Kepler was mathematically difficult, however, and also used obscure epigrams, so he did require expert reading.

On 22 June 1633 [23, 24] Galileo made his final abjuration, in the most abject terms. There is a legend that when he came to the motion of the Earth, he muttered *eppuor si muove*—all the same it moves; but alas there seems no evidence for this. They had broken his spirit, and with a forged document. Who forged it? The Inquisition? It seems unlikely; they did not work that way. One of the witnesses, for money? Possibly, but he would have to have access to the file, which was not kept in Bellarmine's house. It seems most likely that some minor functionary of the Inquisition either through spite or for money, slipped it into the file. This could have been as late as 1632, we have no evidence that it was done in 1616. Whatever the reason, it served its purpose.

Once he had recanted, the sentence was lenient, by the standards of the time. He was sentenced to house arrest, forbidden to publish, and admonished to recite the penitential psalms every week. He returned to Florence and settled back at Arcetri, next door to his daughter's convent. A new blow fell in 1634. His daughter fell ill and died at the age of thirty-three. Though he had treated her badly, which she never resented, she had been his great support in life. Her death was a cruel blow which he only realized when she was gone.

In spite of this and in the face of impending blindness, Galileo set out to write his greatest book, the *Two New Sciences Dialogue*. It is on the foundation of mechanics, which is considered the foundation of all of physics. In it he uses the same three characters as in the *Two Worlds Dialogue*, but this time they are talking about falling bodies, cannonballs, moving ships and so on. He introduces a fundamental change from Aristotelian mechanics. Force is proportional to acceleration, not to velocity. It is the beginning of Newtonian dynamics. He was not allowed to publish, so the manuscript had to be smuggled out and printed in Holland by Elzevir in 1638. In fact this was simple. The guards who were to watch him worked a somewhat casual routine. The Inquisition did not act, although he had broken the terms of his condemnation. By now he was totally blind and chronically ill. He died on 8 January 1642.

The condemnation was a bitter blow, but without it he would probably never have written his greatest work, and Newton would have had nothing to build on. Copernicanism was in some ways a distraction for him. He was a physicist, not an astronomer.

So his sufferings, though painful, were not without product. The real sufferer was the Roman Catholic Church. Some good may have come out of it. The mistake made in the seventeenth century perhaps prevented a worse mistake in the nineteenth, over evolution, where the Church remained silent.

Science was not held up except in Italy. By 1633, everybody with informed knowledge knew that Copernicus was right. Within days of the condemnation the Jesuits were pointing out that the decision was not infallible. It probably did help with the acceptance of Kepler's laws. Throughout Europe, it caused a sensation. John Milton actually hired an agent in Rome to keep him informed. But in Italy it was a tragedy. Science collapsed, and culture in general withered under the shadow of the Inquisition. As a Pope, Urban VIII was a disaster.

Protestant Europe was delighted, though Luther and Melancthon had both condemned Copernicus a century earlier. The consequences lasted for three hundred years, becoming progressively more garbled. The highly respected *London Encyclopedia*, for instance, said in 1829, that Galileo had been imprisoned for some months in 1616, which is quite incorrect; moreover that Galileo's wife had been persuaded by her confessor to hand over important documents for burning. Since Galileo had no wife, this is unexpected. The rationalist press made merry over the affair throughout the nineteenth century. Sometimes Galileo is presented as an early agnostic. This is quite incorrect. He was a devout but anti-clerical Catholic; and this must have increased his agony. Not only his body but his immortal soul was in the hands of the Inquisition.

Progressively he has been rehabilitated. The *Two Worlds Dialogue* was taken off the index of forbidden books in the 1820s. His position on the interpretation of Scripture was vindicated in 1893 by Leo XIII. Finally, in 1992, his condemnation was quashed, after 350 years, but the picture of the Church as narrow, petty and obscurantist will remain.

Curiously enough, the effect on great physicists outside Italy was rather small. French Catholic scientists ignored the papal decrees and continued to work happily with Copernicus and Kepler to guide them. Later on, Isaac Newton, the greatest physicist for a thousand years (until the twentieth century, indeed) has over a hundred references to Galileo in his correspondence. None of them refer to the condemnation. Similarly with Kelvin, and Maxwell. In Hamilton's correspondence, there is one reference, but it is in a letter from de Morgan, and is actually rather favourable to the Church. The encyclopaedists of the eighteenth century actually attacked Galileo for not standing his ground. Easier, perhaps, said than done.

In summary, Galileo was partly his own worst enemy. He was one of the greatest of physicists, but he used arguments which were little better than nonsense, while neglecting his own important discoveries. He became obsessed with the tides, the illusory nature of comets. He was one of the principal founders of the scientific method, but did not always use it himself if a blow at one of his adversaries seemed more fun. However, that he was ill-treated, and in a most contemptible way, there is no doubt. He is an attractive figure, talented not only in science, but also in theology, poetry and literature, and with the gift of making friends. Unfortunately he excelled also at the art of making enemies.

Who finally betrayed him? Or did he betray himself? Was he betrayed at all? Or did he, swollen with pride, finally overreach himself? It is difficult to say. There are elements of all of these, it seems. The Jesuits at the Collegio Romano have been blamed for much of his misfortune, but they did not contribute to the final judgment of the Inquisition. Their failure lay perhaps in not giving him their support in 1610, 1616 and 1633, but as Father Brodrick himself says, they were bound to the support of Aristotle.

It is difficult to find a parallel for the persecution of Galileo; so few people occupying a high position have been so treated for their beliefs. The case of Sir Thomas More, a hundred years before in England, comes to mind. He had been chancellor before his dismissal and execution, having committed no crime. Socrates, perhaps, is a better example.

De Santillana makes an important point in the preface to his book [25]:

> It is difficult to see the actual shape of the conflict in these matters so long as we remain under the spell of a misunderstanding tacitly accepted by both sides: the idea of the scientist as a bold 'freethinker' and 'progressive' facing the static resistance of conservatism. This may well be true on the level of personalities, for it is usually the scientist who shows the freer and more speculative mind in contrast with more prejudiced opponents. But the core of the thing is different: there the scientist appears more often than not as the conservative overtaken by fast-moving social forces. He usually has the Law and the Prophets on his side.
>
> This ought to become immediately clear if we think of contemporary events. [The book was written in 1958.] The tragedy of the geneticists in Russia, with its lamentable apologies and recantations, is a faithful rehearsal of the Galileo story; yet we could not accuse the Soviet Government of clinging to ancient superstitions or of underestimating the pressing need for science and technology. And lest our straining for beams make us overlook our own domestic gnats, we may perceive in the Oppenheimer case a parallel which is a shade too close for comfort.

We will consider the parallels later, in the conclusions in Chapter 13.

Appendix: The validity of the secret minute

The argument of this chapter depends on the postulate that the minute of 26 February 1616 was a forgery, or at least an inaccurate record of what occurred. It is obviously important to validate this claim.

We can say straightaway that the document is at least incomplete. It is not notarized and is not signed by Galileo, which was normally required for an abjuration. It could be a minute for a final document, but if so why is it signed by two servants? Also, if this is so, then Galileo must have been aware over the period 1616–1632, that he was forbidden to discuss the Copernican model even hypothetically. Why did he behave over this time as if he were free to discuss it hypothetically? Why did he risk getting permission from the Pope to write the book? How did he get three papal imprimaturs for the *Two Worlds Dialogue*, one of them from the Master of the Sacred Palace, at the heart of the Inquisition itself, with free access to all the Inquisition files? Why did that same Master of the Sacred Palace complain that the *Dialogue* did not seem very hypothetical?

The strongest argument comes from the text itself. This has been most closely studied by de Santillana [26]:

> Every single legal act, or official letter, was written (or started) on the first verso of a new double sheet and then incorporated

and stitched into the file according to date. This, of course, left a large number of blank second pages in the context. These, too, are numbered. Some of them were used for administrative comment, forwarding notes and follow-up instructions, all in the proper time order. But in this as in any other administration of the times, there is no single letter, report, legal act, or certified copy which does not start on the first page of a new sheet. That is, with one apparent exception: the Bellarmine injunction. This most essential piece is written on space that was only accidentally available, being provided by the back pages of two other documents. Its place as well as its form emphasizes that it is only a minute.

(These pages, folios 378v and 379r, facing each other, are the verso of the blank second page of the Qualifiers' report (folio 377) and the blank recto of what is the second half of p. 357, which belongs to Caccini's deposition. This is the way other transcriptions are set down too; cf those referring also to papal commands, 352v, and the one on an unnumbered sheet following 534. The procedure followed is quite regular as far as the first part is concerned (the Pope's order to Bellarmine on the 25th), for the original of that was supposed to be in the *Decreta* and here only reproduced for information. But then it slips on with deceptive casualness into the second part, dated February 26th, which is the injunction itself and should have been preserved in the original.)

So the final copy, notarized and signed by Galileo, is simply not there. There are no pages cut from the file, apart from two at folio 346 which leave large margins, with no attempt at concealment. Some half sheets are also missing.

The semi-official Jesuit writer, Brodrick [27] comments:

> The certificate, Galileo's own confident attitude and the protocol of the Holy Office issued on March 3, 1616, form a powerful combined proof that the great scientist was at no time called before the Commissary General and forbidden absolutely to hold, teach or defend in any way, either orally or in writing, the theory of Copernicus.

So how do we explain the secret minute? Some degree of sharp practice seems to be indicated.

Stillman Drake has suggested an innocuous explanation. When Bellarmine had concluded his courteous conversation with Galileo, the Dominicans stepped in and pronounced their anathemas. Bellarmine took Galileo aside and told him to take no notice. This would explain why Galileo had no recollection of the injunction, but it is hardly plausible. Galileo was not a doddering old man, he could tell an anathema as well as anyone else. Moreover, it does not explain the irregularities in the document at all. So we reject this explanation.

Perhaps the most interesting problem is what happened to the final document, of which this purports to be a first draft. It should have been stitched into the file. It should have been signed by Galileo. It seems likely that it never existed. But if that is so, why is the minute so amateurish? Was it deliberately intended to be so, thereby explaining why Galileo did not know of it, because it never got to the stage of a final version which had to be signed? If so then why were the witnesses such minor characters? Altogether it strains credulity. We conclude that the minute, if not strictly a forgery, was at least an elaboration designed to mislead.

The physical characteristics have excited attention ever since the anomalous character of the document were pointed out in 1877 by the Catholic historian Reusch. Wohlwill believed that a line had been erased and several lines added, on the basis of an examination with a magnifying glass. Favaro, however, considered that the document had not been tampered with, but could not explain its inconsistencies. Lannert in 1927, subjected it first to soft x-rays, then to ultraviolet light, and he could find no trace of interference. Carbon dating would not be useful in this case.

A slight majority of historians believe that the document was written in 1632, in a (successful) attempt to incriminate Galileo, but the most respected of them believe that it dates from 1616. Who wrote it? That, we will never know. Two people were involved, for the handwriting changes on the second page.

Chapter 4

KEPLER

Johannes Kepler was one of the two or three greatest astronomers of all time, yet he spent his life scraping a living and involved in all kinds of conflict. His greatest work was published only three years before his death. By comparison, Galileo had an easy life.

Kepler was born on 27 December 1571 in Weil der Stadt in Swabia, south Germany. His family was of noble ancestry but in reduced circumstances. His grandfather Sebald Kepler restored the family fortunes somewhat by trade, and became the local mayor. The family was Lutheran, but the city was predominantly Catholic. Johannes' father Heinrich was an unpleasant man, a mercenary soldier. Johannes' mother Katharine was a strange woman, preoccupied with potions and ointments made out of herbs. Neither was popular with the neighbours, they were cantankerous and difficult. Johannes was the eldest but Katharine had six more children, three of whom died in childhood. Johannes himself was a sickly child, who caught smallpox and nearly died. This illness damaged his eyesight, a disaster for a future astronomer in the days before telescopes.

When he was three, his father took off for a war in the Netherlands, followed by his mother, so he had to go and live with his grandfather. His parents returned in 1576 and moved with their family to Leonberg in Württemburg. Heinrich disappeared again but when he returned he opened a tavern, in Ellmmendingen, which failed. Finally he went on another expedition and never came back. There were rumours that he had been killed in a naval engagement in Italy but they were never verified; he just disappeared.

All these moves were bad for Johannes' education, but he was bright and was allowed to enter the Latin school at the age of seven. The Latin schools in Germany were set up at the Reformation. All international diplomatic and administrative correspondence, and most literary work throughout Europe in both Catholic and Protestant countries, was carried out in Latin. Before the Reformation, this work was done by the

Figure 4.1. *Johannes Kepler (1571–1631). Reproduced by permission of the Mary Evans Picture Library.*

monasteries, but after their dissolution there was a need for scholars who would be fluent in written and spoken Latin. The regimen was strict, in these schools only Latin was spoken. He did well but it took him five years instead of the usual three to complete the course. The question then arose as to what he was to do next. His mother had no money, his grandfather was unwilling to part with much and his father had gone. He was not strong enough for farm work, but he was very intelligent. They decided to send him to train for the Lutheran clergy. He passed the entrance examination and was admitted to the seminary at Adelburg. He was recognized as exceptional and passed the entrance examinations of the famous University of Tübingen in 1588. This had a faculty of theology which is still famous today.

In Tübingen Kepler blossomed. He was a very gifted student in all the subjects of the curriculum. For the first two years of the course they studied mathematics, physics and astronomy, before beginning theology. In these Kepler excelled. He also became more outgoing and found friends. In his second year he was granted a small scholarship. For one who had suffered from a poor education as well as poor health it was a notable achievement. More success was to come. He took his master's degree in 1591 at the age of twenty, and was highly placed.

He was given a further scholarship. The Professor of Mathematics at Tübingen was Michael Maestlin. He introduced Kepler to the work of Copernicus [1]. Knowledge of the heliocentric model was limited in Protestant countries because the idea had been ridiculed by Luther on its publication. At this stage he underwent a religious crisis. While not losing his belief in God, he came to doubt some Lutheran doctrines, and was somewhat attracted by Calvinism. Details of doctrine were taken very seriously in the sixteenth century; it was not enough just to describe oneself as a Protestant. Catholic doctrines of course were anathema. At the same time he found himself more and more attracted by astronomy and mathematics, and came to realize that these would be his life's work. However, while prospects were reasonably good in the church, jobs in astronomy were hard to find. He decided to continue towards ordination but hoped that something else would turn up; it did, in 1594.

In Graz, the capital of the Austrian province of Styria, there was a protestant seminary which had a vacancy for a mathematics teacher. This teaching would of course be below university level, but the faculty at Tübingen were asked to nominate someone. They chose Kepler. After agonizing for some days he accepted, always bearing in mind that he might return to Tübingen after a year or two to complete his theology studies. He was now twenty-three. In 1594 he went to take up his position. His salary was small but he was able to supplement it by casting horoscopes for the local nobility. Kepler's attitude to astrology was ambivalent. He did believe that the planets had some effect on human beings, but he was contemptuous of his customers' credulity. In drawing up a horoscope or calendar he made use of his knowledge of politics, military or domestic affairs. Usually the customers were satisfied. He had some spare time, for his subjects were not popular with students. In his second year he had no students at all. He was not a very good teacher; he demanded too much from his pupils, and got carried away in classes with an enthusiasm they could not share. At this time he started to think about the universe in original terms. He has been called the first modern mathematical physicist, and this is true, but he had a mystical, Pythagorean belief in divine harmonies. He asked, why are there six planets (Mercury, Venus, Earth, Mars, Jupiter and Saturn)? Why are they spaced as they are? There must be some fundamental law which determines these things. Now there are of course fundamental laws which determine the behaviour of atoms and subatomic particles, and many interactions are determined by very accurate predictions. The solar system is made up of atoms and these of course obey these quantum laws. The whole planet, however, does not obey laws of this kind, and the modern astronomer would not look for harmonic relations of the kind taught by Pythagoras in looking at the spacing of the orbit or things of this kind. Quantum-type laws with simple integer combinations do not determine such things. Kepler nevertheless was determined to find some.

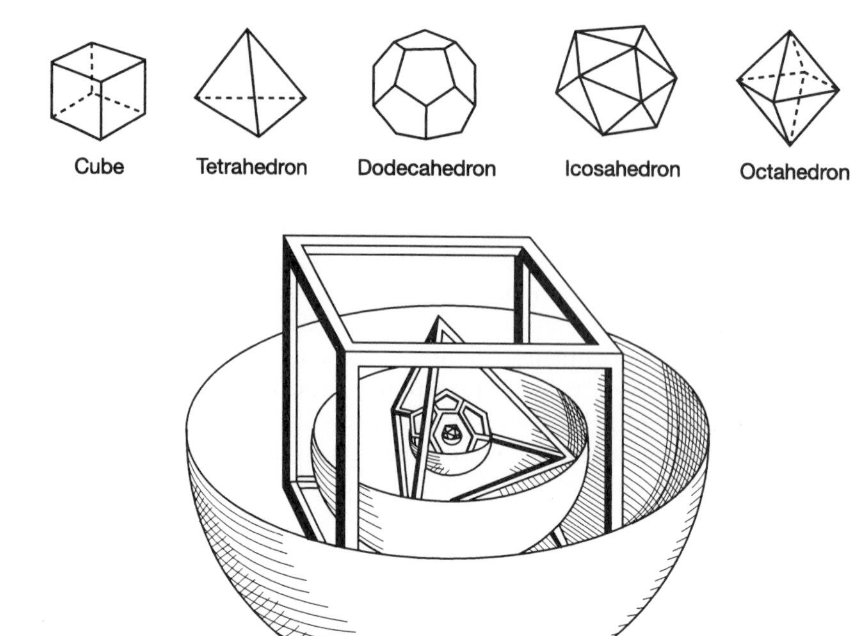

Figure 4.2. *Kepler's Euclidian solids.*

It struck him that there were five and only five regular solids: the octahedron, the icosahedron, the dodecahedron, the tetrahedron and the cube. Euclid had shown that no more are possible. In addition there is the sphere, making six figures. He made a model (figure 4.2). Taking the Earth's orbit as the sphere, he fitted the other solids inside or outside, in the order of the orbits of the planets. It turned out that the relative distances of fitting solids were approximately those of the planetary orbits. We now know that such laws do not apply, and later attempts to account for the planetary orbits in harmonic terms have failed. Kepler was to discover fundamental laws about planetary orbits but this was not one of them. Nevertheless he was filled with a sense of revelation, and published his results in a work called *Mysterium Cosmographicum*, the *Mystery of the Universe*. It appeared in 1597, when Kepler was twenty-six, and caused quite a stir with astronomers and the nobility. The Duke of Württemberg was particularly impressed, and took an interest in Kepler which was useful in his career. The University of Tübingen had agreed to print the book. Michael Maestlin supervised its production since Kepler was still in Graz. It did not make much money but it helped to establish Kepler's reputation. In particular, it fell into the hands of the great Danish astronomer Tycho Brahe. He was impressed, which was important, since the lives of the two men were to become so intertwined.

Brahe was an interesting character as well as being a great astronomer. As a young man he had led a somewhat dissipated life. His nose had been partially cut off in a duel, and he made prosthetic noses in silver and brass, and in gold for special occasions. They frequently fell off. In spite of these eccentricities, he was a favourite of the King of Denmark, who gave him a castle on a small island named Hven. Here he built his observatory, named Uraniborg. It had the best equipment in the known world and with it he made the most accurate observations. For over forty years he carried out observations of the planets, and particularly of Mars. His figures were generally accurate to a minute of arc, but he had no model of the solar system. He did not believe in Copernicus, on both scientific and religious grounds, but he recognized the defects of the Ptolemaic model. So he constructed a model of his own, the Tychonic system. In this, Mercury and Venus revolved around the Sun, and all three revolved around the Earth, with the outer planets. Not surprisingly, it fitted the data well, and preserved the primacy of the Earth. Kepler was a fervent believer in Copernicus, but he revered Tycho as the greatest living astronomer, and he was delighted to get a letter from the great man congratulating him on his book. Tycho also said, however, in a letter to Maestlin, that observations were necessary to confirm Kepler's ideas.

In 1597 Kepler got married. His wife, Barbara, was the daughter of a wealthy merchant, Jobst Muller. She had been married twice before and had a young daughter, Regina, for whom Kepler was a caring stepfather. The next three years were the happiest of Kepler's life. He began to work seriously on the Copernican model, and soon realized that he must have observations. He had neither the equipment—he was a modest schoolmaster—nor the eyesight. Tycho though had both, and a vast stock of unreduced observations built up which he kept to himself. In 1598, however, things changed for the worse.

In 1596 the Archduke Ferdinand, who was ultimately to become Emperor, became ruler of Styria, of which Graz was the capital. He was a Catholic, as were most of the population. The Lutherans were very much in the minority, and the seminary in which Kepler taught served only a small Protestant enclave. Ferdinand began to take measures against the Protestants, first taxes, then an attack on the rights of Lutheran clergy to preach. Finally a decree was issued which insisted that all Protestant teachers leave Styria within two weeks. This, of course, included Kepler. He left but soon was allowed to return. An exception was made in his case because he was district mathematician as well as a teacher. Without him there would be no horoscopes. His position was insecure, however, and he was fined for trying to bury his baby daughter with Protestant rites. So this was his first open clash with authority. Astronomy was not involved, although he was known to believe in Copernicus. Later on that too would be a charge against him.

At about this time Kepler first met Tycho Brahe. The Danish astronomer was having difficulties of his own. His patron the King had died in 1588, and when his young son came to see the observatory at Hven, Tycho insulted him. Since he had insulted many other important people he had no shortage of enemies. His funds were reduced as the years went by until they were almost gone, and he was attacked by hired assassins until he was forced to leave Denmark altogether. He settled briefly in a suburb of Hamburg. The Emperor Rudolf II was a patron of the arts with wide and divergent interests. He heard of Tycho's difficulties and offered him the position of imperial mathematician at a salary of 3000 gulden a year (now worth about £50 000). Accepting it somewhat grudgingly, Tycho arrived in Prague, the imperial capital, in June 1599. He began to set up his instruments, most of which were still on their way from Denmark, at Benatky castle near Prague. He was anxious to get Kepler as his assistant. Equally Kepler was desperate to get Tycho's observations, but could not afford the cost of the journey to Prague. A benefactor offered to take him with him. So the two men met. It must be ranked as one of the great coincidences of history that two men so uniquely qualified to work together should have been at the same place at the same time: Tycho to take the observations, Kepler to analyse them, and to seek the underlying laws. On 4 February 1600, the twenty-nine year old Kepler at last encountered the fifty-four year old Tycho Brahe.

Brahe was conscious that his powers were failing, and it seems possible that he was suffering from the early stages of Alzheimer's disease. He was anxious to have Kepler for the job that he knew he could not do himself. The Emperor wanted a comprehensive set of astronomical tables, known when they finally appeared twenty-seven years later, as the Rudolphine Tables. It was an enormous task. Tycho Brahe died suddenly in November 1601, leaving Kepler greatly distressed. It seemed that he had lost his opportunity to work with the observations, unless he could persuade Tycho's family to part with them. The Emperor solved the problem by appointing him Imperial Mathematician. He brought his family to Prague where prices were high and his wife was unhappy.

Having Tycho's observations he cheerfully set to work. He started with the papers on Mars, a lucky choice as it turned out. The evaluation of the orbit was to take him five years. He was not exclusively concerned with this. He also wrote on refraction, on the supernova of 1604, on mathematical topics and carried on a large correspondence. Mars was his obsession. Applying the Copernican model to Tycho's observations, there were large discrepancies. In fact these arose from the insistence on circular orbits, which were sacred to Ptolemy, Copernicus and also, as we have seen, to Galileo. Kepler too was dedicated to them within the

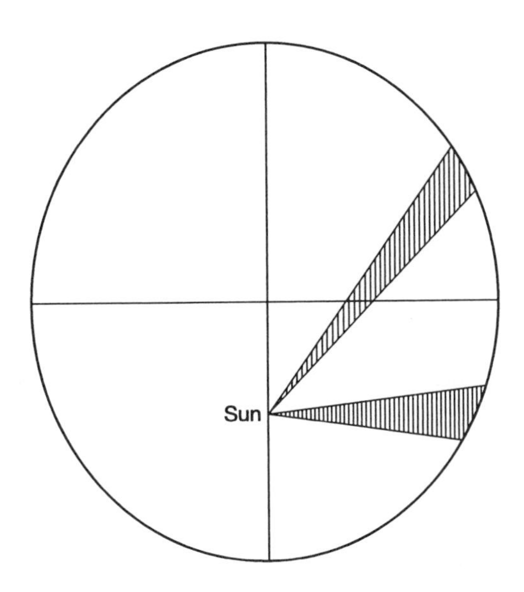

Figure 4.3. *The second of Kepler's laws.*

Copernican model. The errors that were occurring were of the order of eight minutes of arc. This would be acceptable to Copernicus, but Tycho never made errors of this magnitude; about one minute of arc was his typical uncertainty. Kepler experimented with eccentric circles, but these did not work. Then he turned to oval curves, and finally to ellipses [2, 3]. Conic sections—the circle, ellipse, parabola and hyperbola—were not well known in the sixteenth century, and Apollonius' work on them was still the standard. Kepler found it, as Hypatia had twelve hundred years before, very difficult: 'How many mathematicians are there who would put up with the labour of reading through the Conics of Apollonius of Perga?'

He persevered and was rewarded with total success. The orbit of Mars was described within the limits of accuracy, and the others followed. This was the first law of planetary motion, but in fact it had been preceded by another, which we now call the second law. The principle of this is illustrated in figure 4.3. Kepler was working with eccentric circles at this time, in 1602; ellipses did not yet figure in his picture. The law states that for a given planet, equal areas are swept out by the radius vector in equal times. It applies approximately for an eccentric circle and exactly for an ellipse. In the latter case the radius vector is drawn to one of the foci, where the Sun is. (Strictly speaking the planet and the Sun both rotate about the centre of gravity of the solar system which is within the volume of the Sun. It is probably fortunate that this small deviation was not detectable with Tycho's instruments; the theoretical problem would have been very difficult.)

The third law, connecting the orbital period with the size of the orbit, was longer in coming. The second law was discovered in 1602 and the first law in 1605. The third law did not appear until 1618. Although Kepler was Imperial Mathematician, and the Emperor was extremely interested in his work, his salary was constantly in arrears, and much effort went into extracting it from the imperial treasury. Later in his career the treasury owed him 12 000 gulden (£220 000). Also he spent a considerable amount of time casting horoscopes, not only for the Emperor but for the whole court. Rudolf was an unpredictable person and towards the end of his reign in 1612 his sanity was in doubt. In spite of this he was a tolerant monarch in an age of intolerance, and while he did not pay promptly, he was a keen patron of the arts.

4.1. Kepler's conflicts

The tragedies and conflicts in Kepler's life rival those of Bruno in number, but were different in character. Very few of them were directly connected with science, though as a known supporter of Copernicus he was suspect to both Catholics and Protestants. Mainly his problems arose from the Counter-Reformation and the Thirty Years War. We will summarize them, then deal with some of them in detail.

1598	Expelled from Graz by the Catholics, but allowed to return.
1599	Fined for attempting to have his daughter buried with Protestant rites.
1600	Expelled from Graz permanently.
1600	Conflict with Tengnagel over use and ownership of Tycho's observations.
1611	Refused position in Württemburg by the Lutherans. His son Frederick died. His wife Barbara died from typhus.
1612	His patron the Emperor Rudolf died. Living in Linz. Refused communion by the Lutherans.
1619	Appealed to Hafenreffer in Tübingen for permission to work. He is refused.
1618–20	*Handbook of Copernican Astronomy* banned by the Inquisition following the decree of 1616.
1616–20	His mother accused of witchcraft. Kepler defends her.
1621	In Linz. Conflict with both Catholics and Protestants.
1625	His son forcibly baptized as a Catholic. His library sealed up by Counter-Reformation authorities.
1627	Offered a job by the Emperor Ferdinand on condition that he become a Catholic. He refuses.
1629	Persecution by Catholics. Still out of communion with the Lutherans.
1630	Arrives in Regensburg ill. Dies there. Visited by clergymen but not given communion. Imperial salary 12 000 gulden in arrears at his death. His grave is desecrated by soldiers.

Most of these conflicts were not directed against Kepler personally, but arose because he was not a Catholic in a Counter-Reformation environment. His unorthodoxy as a Lutheran even improved his treatment in the Catholic areas. On the other hand, he was refused employment in Protestant Germany because of that unorthodoxy. Copernicanism was only a small part of his heresy; most of his reservations concerned the Eucharist. He was deeply religious but, unusually for his time, extremely liberal in his religion. The Counter-Reformation and the Thirty Years War were disasters for him, mitigated only by his standing as Imperial Mathematician. His excommunication by the Lutherans was a source of great sorrow to him. His expulsions from Graz occurred while he was relatively unimportant, a provincial schoolmaster. He gained some standing because he was district mathematician and was therefore needed to cast horoscopes, but this did not prevent his final expulsion in 1600. In Prague, he was safe for some years at the imperial court, but that too became unsafe. In 1611, occupying troops brought plague and disorder. Kepler's wife Barbara died of typhus, soon after their son Ferdinand died of smallpox. This year too, Kepler was finally rejected in his bid for a job in Tübingen. His theology was unacceptable to the Lutherans, though the Duke of Württemburg would have been glad to have him back. Throughout his life Kepler longed for a faculty position in Tübingen, scene of his happiest days, but he never achieved it.

An opportunity arose in Linz, in Upper Austria, and the new Emperor, Matthias, allowed him to go there, while retaining the post of Imperial Mathematician. It would involve constant travelling to and from Prague but Kepler accepted the job gratefully. He settled in Linz and within two years married again. He spent fourteen years in Linz but was never happy there. He disliked Austria and longed for south Germany. His greatest work, however, was done in Linz.

The years 1616–1620 were the most difficult for Kepler, but also the time of greatest achievement. The first problem arose over his mother. She had always been strange, but as she got old she became very strange indeed. It was inevitable that she should be suspected of witchcraft. As the curse of the Catholic countries was the Inquisition, so the curse of the Protestant countries was the obsession with witches. In Frau Kepler's home village of Leonberg, six women were burnt in one year. She was accused of causing a man to be lame and of driving cattle mad. She had a quarrel with a family called Reinbold who blamed all their misfortunes on her. In various ways they plotted against her until she was arrested and tried. Kepler had been sending off petitions on her behalf, now he travelled to defend her. After fourteen months of imprisonment the Duke ordered that she should be released. She died six months later in 1620. Kepler's young daughter Katharina had also died in 1617.

At this period Kepler published his *Handbook of Copernican Astronomy*, in three parts, appearing in 1618, 1620 and 1621. This was a textbook suitable for use in schools. It was immediately banned by the Inquisition under the edict of 1616 which Galileo had provoked, though operating in a religiously mixed area they could not take direct action against Kepler. In 1618 after long and arduous work, Kepler made his greatest discovery.

He had long been convinced that there must be some relationship between the period of a planet and the size of its orbit, but had been unable to find it. He experimented with harmonic relations by analogy with music, as he had worked with the regular solids twenty years before. Slowly he came to see that the relation which he was looking for was not of this kind. Finally he came on his third law of planetary motion. The square of the planetary year is proportional to the cube of the mean distance from the Sun. This is a strange relationship which can only be understood with the aid of calculus and Newton's theory of gravitation.

> After I had discovered true intervals of the orbits by ceaseless labour over a very long time and with the help of Brahe's observations, finally the true proportion of the periodic times to the proportion of the orbits showed itself to me. On the 8th of March of this year 1618 ... it appeared in my head. But I was unlucky when I inserted it into the calculation and rejected it as false. Finally on May 15, it came again and with a new onset conquered the darkness of my mind, whereat there followed such an excellent agreement between my seventeen years of work at the Tychonic observations and my present deliberation that I at first believed that I had been dreaming and assumed the sought-for in the supporting proofs. (D C Knight *op. cit.* p 139)

He published the third law, buried in a vast mass of metaphysical mysticism and speculation. So much so that very few people understood. The young Englishman Jeremiah Horrocks mastered the first and second laws, and would perhaps have done the same with the third law, but died at the age of twenty-two in 1641. Various authors argued about it and it finally filtered out, so that Newton was aware of it in 1665. It was only with Newton's later work that the law was fully accepted.

4.2. Kepler and Galileo

Galileo could be loutish when he chose, and chose to be in his relation with Kepler, who was seven years younger and in his youth deeply respected the older man. Kepler sent a copy of the *Mysterium Cosmographicum* to Galileo in 1596. Galileo never replied but used it in his lectures. Even when Kepler was made Imperial Mathematician, Galileo showed him no respect. When Galileo discovered Jupiter's moons, Kepler took a deep interest, and later showed that they obeyed the third law. Galileo did not reciprocate. He probably did not understand the third law, buried as it was. Kepler was a far greater mathematician, but could not resist introducing Pythagorean mysticism. Galileo was more down to earth and basically was a physicist not an astronomer. He held as basic dogma that the orbits of the planets were circles, and ignored the evidence that they were not. His attitude caused great pain to Kepler, who was sensitive, perhaps hypersensitive.

4.3. The closing years

With the outbreak of the Thirty Years War in 1618, many thousands suffered, amongst them Kepler and his family in Linz. The war is treated briefly in the appendix at the end of this chapter. Linz was occupied by Bavarian troops loyal to the Emperor Ferdinand. The Counter-Reformation was in full swing, though Ferdinand had retained Kepler as Imperial Mathematician, and Kepler was to suffer directly or indirectly from religious discrimination for the rest of his life.

He began work on the long-neglected *Rudolphine Tables,* which he envisaged as much more than a set of tables of planetary positions; rather a complete compendium of his contributions to Copernican astronomy. The Brahe family disagreed, particularly Tengnagel, Tycho's son-in-law. Kepler's attitude to the observations is summarized in his own words [4]:

> Tycho did what Hipparchus did; it is the foundation of the building. Tycho endured the greatest labour. We cannot all do everything. A Hipparchus needs a Ptolemy who builds up the theory of the other five planets. While Tycho was alive I achieved this: I built up a theory of Mars, so subtle that the calculations completely agree with the observations.

One reason for the delay was that the Englishman, Napier, had published his work on logarithms. Kepler saw immediately that this could simplify the interpretation of the tables considerably. It meant many changes to what he had already done. Eventually he finished them in 1623. Now came the problem of printing. It could not be done in Linz because there was no suitable press. Tired beyond his years, on an old and tired mare, the astronomer set out to find a printer. In the middle of war it was foolhardy. Only his faith in the value of what he had done sustained him. It was a spiritual faith rather than a modern scientific agnostic one [5]. There is little like it in the history of science. Ulm seemed a possible place: there were printers and he had friends there. He had run out of imperial funds and had to pay for the paper himself. However, Ulm fell through because the Emperor forbade it. He returned to Linz with a press capable of printing the tables. He travelled to three cities in great danger, to collect the funds allocated by the Emperor. In 1625 it seemed that all was ready, but religion intervened. There was a general expulsion of all Protestants from Linz. Because the Emperor was anxious to get the tables printed, Kepler and the printer were exempted, but Kepler's library was sealed up, and he had to have his baby son baptized as a Catholic. The printing started at the beginning of 1626. In the spring, however, rebellious peasants besieged Linz and the press was destroyed in a fire. The manuscript survived.

The siege was lifted in mid-1626 and the Emperor agreed to let Kepler go to Ulm to print the tables. He decided to leave Linz altogether. In November 1626 Kepler and all his family left Linz by boat on the Danube. He reached Ulm but not by boat. The Danube froze and he had to leave his family in Regensburg, travelling on himself by road. In Ulm he found a printer but the man turned out to be incompetent and dishonest. He decided to try in Tübingen. For him, Tübingen was always better. On the road, however, he became ill and had to return to Ulm.

The printing of the *Rudolphine Tables* was completed in September 1627. Kepler took some copies to Frankfurt to the famous book fair, which still exists. The Tables became a standard work for astronomers, astrologers and navigators, but initial sales were small. The Brahe family were also annoyed because they felt they had not been given their share of the credit, and insisted on changes in the second edition. Returning to Prague to present a copy to the Emperor, Kepler was very well received and was offered an attractive appointment. The snag was that he had to become a Catholic. This he refused in spite of strong pressures from all sides; the Emperor was not pleased. Kepler's future looked in doubt, but at this time another patron appeared, the flamboyant General Wallenstein. Kepler had cast a horoscope for him many years before and it had turned out to be accurate. He took up the offer and left for a new home in Upper Silesia, at the town of Sagan. This was a Protestant town and he was free from harassment on religious grounds. Sagan is on the Oder river in northern Germany. Kepler was homesick for southern Germany and particularly for Tübingen which had always rejected him. When not drawing up horoscopes for Wallenstein he worked on an ephemeris. Wallenstein paid him 1000 gulden a year, less than the imperial salary, but it was paid promptly. The Emperor had ordered that Wallenstein should also pay the 12 000 gulden of arrears. This was not forthcoming. One of the advantages of Sagan was that the Counter-Reformation had not reached it, but it did in 1629. Again Kepler was exempted but Protestants in general had to leave.

His family life was going well. His wife had a baby in the spring of 1630, and his daughter was married at about the same time. He had great difficulty over money, however, and decided to go and look for his 12 000 gulden from the Emperor. He was also bound for Linz where he hoped to cash in some bonds. He was deeply depressed and older than his fifty-nine years. His long separation from the Lutheran church was preying on his mind. He reached Leipzig and stayed with a friend, then travelled on to Regensburg. On 2 November 1630, he sold his decrepit horse for 2 gulden, and feeling fever coming on, took to his bed. The

Emperor heard about it and sent him money but far less than he was owed. It was little use to him. He did not respond to the surgeons' treatment. Clergymen visited him but did not offer him communion. He died on 15 November 1630. He had a splendid funeral, but far to the north in Sagan his family were half-starved as he had been. He was buried in Regensburg but his grave was later destroyed in the course of the Thirty Years War.

So, after a life struggle against petty persecution, and of great courtesy and gentleness, Kepler died. He is a more attractive figure than the other two giants of the seventeenth century, Galileo and Newton. His conflicts were lifelong but overtly religious, though his support for Copernicus did not help. The condemnation of his *Handbook of Copernican Astronomy* was the only directly scientific persecution and it came from the Catholic side. The refusal of the Lutherans to admit him to the sacraments over thirty-five years was perhaps more hurtful. He was a man of opposites, as Marie Boas has said [6]:

> ... mystic and rational, mathematical and quasi empirical, he constantly transformed apparently metaphysical nonsense into astronomical relationships of the utmost importance and originality. Immensely arrogant in his conviction that he held a sure key to the mysteries of the universe, and even to the structure planned by God at the Creation, he always acknowledged his debt to his predecessors.

Kepler believed that he was enlightened by a guiding spirit, which he called his genius (not to be understood in the modern sense). The guiding spirit has gone but the belief in the beautiful and sublime has survived. Dirac, one of the most respected physicists of this century, constantly stressed mathematical elegance as a criterion of validity. Current theories of sub-nuclear particles illustrate this belief [7].

Appendix: The political background

The period between the birth of Copernicus (1473) and the death of Galileo (1642) was one of the most eventful in European history. It saw the development of printing, the discovery of America, the Reformation and Counter-Reformation, and at the end of the period, the horrors of the Thirty Years War. Each of these influenced the growth of astronomy and of astronomical controversy.

Printing released a flood of cheap books into society. Greek authors on the physical sciences and astronomical writers such as Archimedes, Aristarchus, Ptolemy and Eratosthenes became freely available. Thus Copernicus knew that Aristarchus had argued that the Earth rotates about the Sun, and that the stars were very distant; and he knew the planetary model of Ptolemy in great detail. The Renaissance, at its height in the fifteenth century, concentrated on the arts and humanities, but science came along with them.

The Reformation and Counter-Reformation gave rise to religious fundamentalism in contrast to what had occurred in the previous century. This was true of Lutheranism, Calvinism and Catholicism. Luther and others condemned Copernicus before the Catholics condemned Galileo. They also caused barriers between countries which had not been there before, and this was inimical to the free exchange which is so necessary to the progress of science.

The discovery of America brought a vast wealth into Europe, particularly into Spain, where it was frittered away in the early seventeenth century, but it also enriched royal courts throughout Europe. Monarchs now could afford to employ many retainers, including astronomers, whose official duties were to cast horoscopes, but who also had time for serious astronomy. Tycho Brahe and Kepler were both imperial court mathematicians, and made their living in this way, though Kepler's salary was at one time three years in arrears, under the Emperor Rudolph. This wave of prosperity was, however destroyed by the Thirty Years War, which broke out in 1618. Before we consider this we must say something about individual states and authorities in Europe.

The Empire

The tenuous descendant of the Holy Roman Empire of Charlemagne had come under the control of the Hapsburg family, though it was electoral. There were seven electors, mainly German. In 1620 there were three Protestant and four Catholic electors, so that the imperial succession was always in some jeopardy. The Emperor Charles V had abdicated in 1555, and the imperial lands were divided. Charles' brother succeeded to the Empire. Charles' son succeeded to the throne of Spain as Philip II, one of the most effective kings of that country. The imperial lands had been very extensive, but by the seventeenth century were reduced to Austria, Bohemia, Hungary (threatened by the Turks) and some territory on the eastern coast of the Adriatic. Because of its hereditary ties with Spain, the Empire also influenced that country with its possessions Portugal and southern Italy. The Hapsburg Emperors, of course, were Catholic. The Empire had some moral authority over the many states of Germany, but could not enforce it, particularly in Lutheran and Calvinist states. The Lutherans and Calvinists were also seriously divided from each other.

The Papacy

In the sixteenth and seventeenth centuries the papal states were not large compared with France or Spain, but they were strategically placed, cutting Italy in half, and denying movement between southern Italy and the north. A significant part of papal income was derived from customs tariffs on goods in transit. The Papacy was in decline during the late fifteenth century and early sixteenth century but later recovered somewhat. The popes were mostly mediocre or plain bad, but as rulers of a state had some secular influence in addition to the religious authority which they exerted over Catholic states. The Augsburg agreement, known as *cuius regio*, stated that the religion of the people of a state should follow the religion of the prince; this worked fairly well, though there were serious problems in mixed areas like Bohemia, and these contributed to the war which was to come.

France

There was a militant and aggressive Huguenot (Calvinist) minority, and the greatest French king of the sixteenth century, Henri IV (reigned 1589–1610) was brought up a Calvinist. He converted to Catholicism during his reign, but during his reign and later, France tended to align itself with the Protestant states. This was so even under Cardinals Richelieu and Mazarin, though they fought internal wars against the Huguenots. French policy in the seventeenth century was complex. It was a wealthy country though vast sums were squandered. It acted as paymaster for countries such as Sweden, which was Protestant under Gustavus Adolphus.

Spain

After the death of Philip II (1598), who left his country demoralized, divided and bankrupt in spite of American revenues, Spain was in a very unstable condition. This arose largely from Philip's incessant pursuit of heretics within and beyond his kingdom: unsuccessful wars against the Turks, England and the Dutch. Subsequent kings were no better. Thus while aligned with the Emperor by blood ties and religion, the country was unable to assert the position which it should have held in the struggle that was coming.

Venice

The republic was past its highest point as a commercial, maritime and military power by the beginning of the seventeenth century, but it was still a centre of independent thought. This is not to say that it was a democracy in the modern sense. Ogg says [8]:

> The truth is that such a vague term as democratic can be used of seventeenth century organisations only with very careful reservations. The political heritage of the Renaissance was absolutism in the state.

Relations with the Empire were somewhat tense. In order to resist the encroachment of the Turks on imperial territory, the Emperor had imported a number of Bosnians into the north-eastern Adriatic. These turned to piracy, attacking Venetian shipping and seriously hampering trade.

Relations with the Papacy were also bad. They revolved mainly over the question of papal authority over sovereign states. This culminated in 1606 when Pope Paul V excommunicated the Venetian Senate and put the country under interdict. The Venetian clergy took no notice, except for the Jesuits, who were expelled and not allowed back into Venice until 1657.

Germany

Germany was a geographical term in the seventeenth century, rather than a political one. There were 350 separate political entities in the area, each with its foreign policy and diplomatic relations. Only about half a dozen were large enough to really count in international politics, but the future of Europe lay with the small princely states.

The Thirty Years War

The war started in 1619 with a rebellion against the Empire in Bohemia. It spread to the German states and Denmark on a roughly religious basis; Catholic versus Protestant. This phase lasted ten years. Ogg says [9]:

> The first decade of the Thirty Years War witnessed the disappearance of Bohemia from European nationalities, and of the Palatinate from the German Protestant states; the futile intervention of Denmark; the Catholic resumption of secularised bishoprics, and most important of all, the ascendancy of the Empire at the expense of Protestant Germany, culminating in the menace of a powerful and reactionary Empire.

This was followed, however, by the intervention of Sweden, subsidized by France, on the Protestant side. In 1629 the French, led by Richelieu himself, crossed the Alps and captured the strategic fortress of Pinerolo in Savoy. Under imperial supervision this was handed back in 1631, but immediately re-captured by the French, following a secret deal with the Duke of Savoy. From now on the French took a more active part in the war.

From 1631 until 1648 the character of the war changed from a primarily religious and territorial one to a dynastic struggle. Civilian deaths from murder, starvation and disease were enormous, not to be surpassed until the twentieth century. The war was finally settled, for the time being, by the treaty of Westphalia in 1648. In this, the Empire, France, Sweden, Brandenburg and Bavaria all made small gains. France acquired Alsace, a doubtful asset which was to cause enormous trouble two hundred years later. Brandenburg made large gains in the East, later in the century. This was the nucleus of the future Prussian Empire.

As far as Italy was concerned, Cosimo of Tuscany sent troops to help the Empire. Pope Urban VIII pursued a devious path, ostensibly supporting the Catholic Emperor, but plotting with the French and the Swedes, and for good measure offering himself as a mediator. Florence was not fought over, and Galileo was affected only indirectly, but his work, during his most creative years, was a trivial side issue in the convulsion that was shaking Europe, as far as the ecclesiastical administrators were concerned.

The centenary of Luther's manifesto at Wittenburg occurred in 1617. This provoked a flood of theological works of a polemical character on all sides. Bellarmine was concerned with these, and it may be that his judgment in the conversations with Galileo was somewhat affected.

England

England was little affected by the Thirty Years War, but had of course its own troubles with the struggle between King and Parliament. Small bands of English semi-mercenaries fought on both sides on the continent. There were dynastic elements in royal policy in the arrangement of marriages.

Chapter 5

INTERNAL CONFLICT. THE ATOMIC THEORY AND POSITIVIST PHILOSOPHY

5.1. The people

The nineteenth century saw enormous advances in physics, though not comparable, perhaps, to those made in the twentieth century. Two important and related fields were the kinetic theory of gases and thermodynamics. Though developed together, they separated late in the century and were used to create a split in physics. This split, its consequences and the people involved are the subjects of this chapter. Looking at fuzzy pictures of bearded Victorian physicists, they all seem rather similar, but they had widely different temperaments. We list some of them in chronological order.

James Prescott Joule was born in Salford, Lancashire, in 1818, the son of a wealthy brewer (the firm still exists). He studied under John Dalton, but was largely self-taught. His most important contribution lay in showing that heat was a form of energy, and in finding the relation between the two units used to measure them. He also put forward a primitive model of the molecules in a gas and showed that it would account for some of the observed properties. He died in 1878.

Rudolf Clausius was born in 1822, in Koslin, Prussia. He died in Bonn in 1888. He tended to work in an applied or engineering environment, as is shown by his appointments: the Artillery and Engineering School, Berlin, 1850; Zurich Polytechnic 1855; University of Würzburg 1867; University of Bonn 1869. He contributed a great deal to the early development of kinetic theory, however, sharing with Joule in the development of simple but useful models.

Figure 5.1. *William Thomson, Lord Kelvin (1824–1907). Reproduced by permission of the Mary Evans Picture Library.*

William Thomson, Lord Kelvin was born in Belfast in 1824 and died in Largs, Scotland in 1907. He is one of the most important figures in the development of kinetic theory and in a wide variety of other fields, both basic and applied. His father was a teacher of mathematics, who got the Chair of Mathematics in Glasgow when Kelvin was a child, so the family moved there from Belfast. His mother died when he was six, and he developed a close relationship with his father. He was the fourth child out of seven. We will consider his contributions to kinetic theory in some detail later, but it is worth saying something about his applied work. He was closely associated with the first trans-Atlantic telegraph cable, risking his life several times during the cable laying. He was knighted for this. He invented the mirror galvanometer for use with the cable. He also invented methods of correcting magnetic compasses in iron ships, and many other things. He was educated first privately by his father, then at Glasgow University (where he entered at the age of ten), from where he went to Peterhouse, Cambridge and then studied in Paris before being appointed Professor of Physics in Glasgow in 1846 at the age of twenty-two. He stayed in Glasgow until his retirement, and also had a successful commercial company. He was created a baron in 1892, and is buried in Westminster Abbey next to Isaac Newton, such was his popular following.

Figure 5.2. *James Clerk Maxwell (1831–1879).*

James Clerk Maxwell was born in Edinburgh in 1831 and died in Parton, Scotland in 1879. He was probably the greatest physicist of the nineteenth century, a greater one than Kelvin, but without the enormous breadth of practical and commercial interests. He was primarily a theorist, but was also a competent and inventive experimentalist. This is perhaps surprising for as a youth he was awkward and gangling and was known to his schoolmates as 'Dafty'. Like Kelvin, he lost his mother at an early age; in his case when he was nine. His father was a successful lawyer and had a small estate. James entered Edinburgh University and went from there to Cambridge. He was appointed Professor in Marischal College, Aberdeen in 1856, but in 1860 was made redundant due to a merger of colleges. He moved to a chair in King's College, London, in the same year. In 1865 he moved home to Scotland and continued his research privately, but in 1871 he was appointed first Cavendish Professor at Cambridge. When he died in 1879 he was buried quietly at Parton with none of the fuss which accompanied the burial of Kelvin. As a physicist he was astonishingly versatile. His greatest work, which endures today, was on electromagnetism, but as the leading spirit in kinetic theory in its maturity he has great historical importance. He had an engaging personality and wrote poetry and satirical verse. His death at the age of forty-eight was an enormous loss.

Figure 5.3. *Ernst Mach (1838–1916).*

Ernst Mach was born in Turas, Moravia in 1838. He took his doctorate at Vienna in 1860 and was appointed to a mathematics chair at Graz in 1864. From there he moved to a physics chair in Prague in 1867 at the Charles University. In 1895 he went back to Vienna as a Professor of Physics. He suffered a stroke in 1901 but remained active until his death in Vienna in 1916. He did important work on the flow of fluids at high speeds, and is remembered in the 'Mach number', but in science he is equally remembered for his positivist philosophy, which influenced whole generations of physicists and chemists. In a sense, he is the villain of our story, together with Wilhelm Ostwald.

Ludwig Eduard Boltzmann was born in 1844 in Vienna. He took his doctorate in Vienna, and held professorships in Graz, Munich and Leipzig before returning to Vienna as a professor in 1895. He was a highly strung, nervous person, deeply dedicated to his work. He suffered from bouts of depression and died, probably by suicide, in Duino, Italy in 1906. He was the leading figure in the second phase of kinetic theory, when it was put on a sound mathematical basis and broadened into statistical mechanics. In the later stages of his work he was in opposition to Mach, Ostwald and other positivists over the existence of atoms. He felt this controversy deeply. He is the central personality in this chapter.

Friedrich Wilhelm Ostwald was born in 1853 in Riga and died at Leipzig

Figure 5.4. *Ludwig Eduard Boltzmann (1844–1906).*

in 1932. He was awarded the Nobel Prize for Chemistry in 1909. He took his doctorate at Tartu in 1878, and taught in Riga before going to Leipzig. He did important work on catalysis, but his influence in physics was probably bad for the subject. Like Mach he was a positivist.

Josia Willard Gibbs was born in 1839 and died in 1903. He inherited a modest fortune from his parents, who died early. He entered Yale in 1854 and worked first in engineering. Having some money, he was free to travel, and spent three years in Europe. In 1871 he was appointed Professor of Mathematical Physics at Yale. Like Boltzmann, he was a most important figure in the later phase of kinetic theory.

Max Planck was born in 1858 in Kiel and studied at Munich and in Berlin. He succeeded Kirchoff at Berlin University in 1889. While best known for his quantum theory, he also did important work in thermodynamics. He was a most respected figure in German physics, and Boltzmann was anxious for his support. Personally, he had a tragic life; one of his sons was killed during the First World War, and another was executed by the Nazis after the Hitler bomb plot of July 1944.

5.2. Background to the conflict: determining the size of an atom

The controversy over the atomic constitution of matter is one of the most long-standing ones in physics, originating with Leucippus in about

500 BC and Democritus in about 400 BC and finally terminating soon after the death of Ludwig Boltzmann in 1906 AD. In its early days it was associated with atheism, and opposed by Aristotle. Epicurus, however, incorporated it into his cosmology, which was essentially anti-religious. Lucretius also dealt with the theory in his poem *De Rerum Natura*, which is a defence of Epicureanism against the Stoics. Since then the question of atoms has been the cause of many conflicts in the history of science. The present chapter will be mainly concerned with the one at the end of the nineteenth century between the positivists and the supporters of the kinetic theory of gases.

The Epicurean atom was indivisible and devoid of properties except impermeability and shape. Weight was not considered. By and large the Epicureans had a bad press in the Middle Ages, mainly because of their opponents, the Stoics, and because of their alleged atheism. This was not quite fair, as was recognized by Thomas Aquinas. The atomic hypothesis deeply influenced such people as Roger Bacon, but most people felt uneasy about it.

Atomism in the sixteenth and seventeenth centuries became important because of the controversies over Eucharistic theology. Giordano Bruno considered the monad, which is closely related. The *Oxford English Dictionary* describes the monad as 'A unit, the ultimate unit of Being (e.g. a soul, a person, an atom, God)'. Problems arose with the classical monad. If it was finite in size, it must, at least in principle, be divisible, but then it would not be an atom. Other problems arose because of the very different properties of gases, liquids and solids of the same material. The very simple picture which kinetic theory introduces did not seem to have emerged until the mid-eighteenth century. Before this, in the seventeenth century, there had been some progress. The properties of gases were investigated by Richard Towneley, a Lancashire gentleman who, because of his Roman Catholic religion, was excluded from Fellowship of the Royal Society, even though he succeeded in mixing freely at their meetings. The true introduction of the corpuscular philosophy into science, however, was made by Boyle. Later on, Newton was interested in atoms, but not primarily in their behaviour in gases.

Towneley's achievement, generously acknowledged by Boyle, was to show that 'Boyle's Law' applied for expanded as well as compressed gases. This was of primary importance, it demonstrated the presence of a fundamental law, but though Towneley and Boyle may have had their suspicions, it did not explain the different behaviour of solids, liquids and gases. The answer was simple, and due mainly to Daniel Bernoulli in the following century. Atoms are small compared with the spaces between them in gases. There are forces exerted on collision, but these are short-range and do not affect the situation very much at large distances. For solids and liquids, however, the atoms are packed close together, and short-range forces are important. These are Descartes' vortices and

Newton' short-range forces. For a clear consideration of low-pressure gases it was necessary to wait for Maxwell and Kelvin.

In Aberdeen, in the mid-nineteenth century, there were two colleges. By order of the University, it was decided to have only one. Therefore it was necessary to dispense with the services of one professor in each subject. In physics the unfortunate person who had to look for employment elsewhere was James Clerk Maxwell. Although he was undoubtedly the greatest physicist of the nineteenth century we should not be too hard on the Electors of Aberdeen. He had not yet shown his full potential even though his understanding was far beyond that of ordinary men. His greatest contribution was in electromagnetism, which remains unsurpassed, but his contribution to the kinetic theory of gases is perhaps of comparable historical importance. For the first time, he, Loschmidt and Kelvin showed that it was possible to determine the sizes and other properties of atoms from laboratory-scale experiments. This was after 2500 years of speculation, which began with Leucippus and Democritus, and it is surprising that for at least fifty years before Maxwell, the materials had been to hand.

Brush says [1]:

> James Clerk Maxwell read Clausius' paper in English translation in 1859 and figured out a clever way to refute the kinetic theory by deducing from it a falsifiable consequence: the viscosity of a gas of tiny billiard balls must be independent of density and must increase with temperature. Everyone knows that a fluid flows more slowly as it gets thicker and colder, but since common experience does not necessarily extend to gases, Maxwell asked G G Stokes, the expert on hydrodynamics, if any experiments had been done on this point. He was presumably reassured by Stokes that experiments on the damping of pendulum swings in air proved that the viscosity of a gas goes to zero as its density goes to zero; in any case he wrote that 'the only experiment I have met with on the subject does not seem to confirm' the prediction that viscosity is independent of density...
>
> The year 1865 [Maxwell was thirty-four] saw the triumph of the kinetic theory. Maxwell found that the viscosity of air is indeed constant over a wide range of densities. This result was fully confirmed by other physicists. It turned out that what Stokes had claimed to be an experimental fact was not a fact at all but a theory-laden observation: in reducing the data it had been assumed that the viscosity of air goes to zero as the density goes to zero. It seems obvious that if there is no air it cannot exert any viscous resisting force. Common sense, or a principle of continuity in nature, would suggest that if there is a very small amount of air the viscosity must be very small. But the actual

pressure range in which this behaviour occurs happens to be below that attainable in the first half of the nineteenth century, so the presumed continuous decrease of viscosity to zero was not directly observed.

The enormous importance attached to this result cannot perhaps be entirely justified. Experimentally, the viscosity certainly is nearly constant from about a thousandth of an atmosphere to about fifty atmospheres. But the theory which predicted this was full of radical simplifying assumptions. In particular it was assumed that all molecules crossing a plane came from a collision plane one mean free path away [2]. A full theory taking into account the distribution of collision points is too complicated for the elementary level at which kinetic theory is usually taught to undergraduates. It gives the same answer as the simple theory over most working pressures.

In 1865 Loschmidt began a process that resulted in the calculation of the size of an atom. He showed that it was possible, from the viscosity formula, to derive the size of atoms, provided a second independent equation can be found. The expression for the coefficient of viscosity is $A = nmLc/3$, where A is the coefficient, n is the number of molecules/unit volume, m is the mass of a molecule, L the mean free path and c the mean molecular velocity. A is measurable and the product nm is the macroscopic density. c or something very close to it, can be deduced from Boyle's Law, as Clausius pointed out in a classic paper in 1857. Therefore, L can be determined.

Once L is known, a further equation is required: $L = 1/na^2$, where a is the molecular diameter so we must eliminate n. The volume of the liquid, where the molecules are close-packed, is roughly na^3. Hence n drops out and a can be found. The value that Loschmidt derived was about four times too big, but within the order of magnitude of the modern value. This was the first estimate of the size of an atom in the 2500 years since Leucippus and Democritus.

The kinetic theory also predicted that the coefficient of thermal conductivity should be independent of gas pressure, down to pressures where the mean free path is becoming comparable to the dimensions of the container. The kinetic theory is found to be true over a wide range of pressures. Thermos flasks must have a very high vacuum to operate. A rough vacuum is not sufficient.

The theory also predicts that the variations of both viscosity and thermal conductivity with temperature are proportional to the square root of the absolute temperature. When these predictions were tested experimentally the viscosity and conductivity appeared to be more accurately proportional to the temperature itself. Much effort went into refining the model of the molecular collisions, taking into account the persistence of velocity after collision, and other factors. In fact, when the measurements are extended to higher temperatures, up to 1500 degrees absolute, the variation is found to be approximately a square root, as predicted by the simple theory.

In the same year, 1865, Maxwell was preoccupied with the distribution of molecular velocities in gases and the related problems. Kelvin, a more pragmatic character altogether, concentrated on the size of molecules by a different method. In 1870 he presented an argument based on the latent heat of fusion and the surface tension. This gave an answer very similar to Loschmidt's value [3].

Daniel Bernoulli, around 1740, had realized that if the sizes of atoms became, by compression, comparable to the separation between them, then Boyle's Law would no longer hold. This was developed by Andrews, and by van der Waals, towards the end of the following century. Plotting pressure against volume, the ideal Boyle's Law should be a set of rectangular hyperbolae. At high pressures, however, these flatten out, eventually, with liquefaction, becoming horizontal straight lines. This was important evidence not only for the atomic nature of matter, but also for the finite size of atoms, and for the forces between atoms. Indeed, the existence of atoms and molecules was reliably established by about 1870.

Even in 1865, however, the problem of specific heats was in the background. As time went on it became more important until it dominated discussion. The basic ideas are not complicated. The specific heat of a body is the amount of energy which must be put in to raise the temperature by 1 °C. For a gas there are two specific heats. We can fix the volume when we heat it by means of a rigid container so that it cannot expand. This gives the specific heat at constant volume, C_v. Or, we can allow the gas to expand when heated, so that it remains at constant pressure. This gives the specific heat at constant pressure, C_p. C_p is larger than C_v because when the gas expands it does work against the environment, and the energy must be supplied by the heat source. The ratio C_p/C_v is an important constant for a given gas. It can be determined experimentally by a number of methods, the simplest perhaps being a determination of the speed of sound in the gas. This depends on the ratio C_p/C_v.

Kinetic theory can be used to make a theoretical prediction of the ratio. A fundamental theorem of wide application outside kinetic theory as well as within it is the equipartition of energy. A molecule can have several degrees of freedom—movement in the three perpendicular directions, rotation about three perpendicular axes and possibly internal vibration. The equipartition of energy states that the average energy in each degree of freedom will be the same.

A monatomic molecule will have three degrees of freedom only; a diatomic molecule should have six, three translational and three rotational. It could also have vibrational degrees of freedom. Similarly a polyatomic molecule will have six or more.

Experimental measurements on common diatomic gases such as nitrogen, however, indicate only five active degrees of freedom, and a ratio of 1.4 for C_p/C_v when 1.33 was expected. This was known in about 1860, but not explained until 1908.

By 1895, Ernst Mach and Wilhelm Ostwald had launched a campaign against the existence of atoms. They fastened particularly on the work of Ludwig Boltzmann. Using singularly unpleasant tactics, they sought to prove that only energy levels can be afforded real status; the concept of an atom is meaningless. Boltzmann, thus attacked, sought the support of his physicist colleagues against the philosophers and chemists. He did not get it. In particular he did not get the support of Max Planck, for some reason which is not clear, since Planck professed to be completely convinced by the atomic hypothesis.

In 1902 Boltzmann attempted suicide; in 1906 he succeeded.

In retrospect we are facing a ridiculous situation. By 1906, Einstein's paper on Brownian motion was published; radioactivity, which implies atoms, was well investigated. All the evidence was there from kinetic theory. The scientific world was on Boltzmann's side. Nevertheless, Mach was nominated for a Nobel Prize in 1909. In 1908 Planck launched a violent attack on Mach and the positivists, but it was too late for Boltzmann.

In analysing the conflict between 1895 and Boltzmann's death in 1906, we are faced with a number of paradoxes. An unholy alliance between the positivist philosophers and the German chemists sought to undo what had in fact been partly done by chemists. In the late eighteenth century and the early nineteenth, most of the important advances came from such people as Scheele, Lavoisier, Priestley and Dalton. Though their atomism was, if such is possible, qualitative, it laid the groundwork for the efforts of Maxwell, Kelvin and Boltzmann. Outside Germany most chemists were committed to atomism; but German chemistry was very influential.

Since 1900, we had had quantum theory, but this was resisted. (Russell MacCormach's documentary novel *Night Thoughts of a Classical Physicist* draws a very accurate picture of a physicist in 1918, scarcely touched by either quantum theory or relativity, but conscious nevertheless of the collapse of his world.) This conservatism was reflected in the Nobel Prizes. Planck was not awarded his until 1918 for quantum theory. (Einstein had to wait even longer, until 1921; and then he got it for the photoelectric effect, not for relativity.) The politics of Nobel Prizes, particularly before the First World War, was strange. Rutherford, for instance, got a Noble Prize for Chemistry but not for Physics. In 1912 Dalen got a Noble Prize for Physics for inventing an automatic light-buoy, an ingenious device, but not, one would have thought, worthy of a Nobel Prize, though here there was an embarrassment. It was proposed to honour Tesla and Edison jointly, but

Tesla refused to share the prize with Edison, so Dalen got it by default. All three were inventors, of course, rather than physicists, but this was not regarded as an impediment.

Why did Boltzmann commit suicide? He was born in 1844, so in 1906 was sixty-two. His major work was completed in his late twenties and early thirties, so he was an established figure. Indeed he could have been a candidate for a Nobel Prize, but this was still very new, originating only in 1901. He held the prestigious Chair of Physics in Vienna. Altogether his power and influence should have exceeded that of his antagonists and, amongst physicists, probably did, but he desperately needed the support of Max Planck. This for some reason was not forthcoming when it was needed, and when it did come, it came, as we have seen, too late. Perrin's experimental studies of Brownian motion established the existence of atoms beyond all reasonable doubt, as even Ostwald conceded, but that was in 1912.

Brush says [4]:

> One of the two clouds over the dynamical theory of heat and light at the end of the nineteenth century, according to Lord Kelvin (1901), was the discrepancy between experimental specific heats and the predictions of kinetic theory based on the equipartition theorem.

Later Brush writes [5]:

> Gibbs avoided the specific heat problem by abstaining from detailed calculations for special molecular models. In the preface of his *Elementary Principles* he gives his rationale, which needs to be quoted at somewhat greater length than usual because one or two sentences taken out of context have sometimes been used to give the impression that Gibbs was an anti-atomist in the camp of energetics or positivism.
>
> ... we avoid the gravest difficulties when, giving up the attempt to frame hypotheses concerning the constitution of material bodies, we pursue statistical inquiries as a branch of rational mechanics. In the present state of science, it seems hardly possible to frame a dynamic theory of molecular action which shall embrace the phenomena of thermodynamics, of radiation, and of the electrical manifestations which accompany the union of atoms. Yet any theory is obviously inadequate which does not take account of all these phenomena. Even if we confine our attention to the phenomena distinctively thermodynamic, we do not escape difficulties in as simple a matter as the number of degrees of freedom of a diatomic gas. It is well known that while theory would assign to the gas six degrees of freedom per molecule, in our experiments on specific heat we cannot account for more than

> five. Certainly, one is building on an insecure foundation, who rests his work on hypotheses concerning the constitution of matter.
>
> Difficulties of this kind have deterred the author from attempting to explain the mysteries of nature, and have forced him to be contented with the modest aim of deducing some of the more obvious propositions relating to the statistical branch of mechanics...

So Gibbs believed in atoms but was not prepared to speculate about them, or to build models. It is not a positivist view, but is sometimes presented as such. The number of positivists was in fact never very large in the nineteenth century. Only Mach and a few others believed that macroscopic energy levels were the only reality.

5.3. Kinetic theory, macroscopic thermodynamics and positivist philosophy

Both the kinetic theory of gases and macroscopic thermodynamics had respectable pre-histories before they took off in the mid-1850s. But during the latter half of the nineteenth century kinetic theory and thermodynamics diverged, and, at least for some scientists, became bitterly opposed. Maxwell, Planck, Clausius, Kelvin, Helmholtz and even Boltzmann worked in both fields, but under the authority of the chemist Ostwald, and the philosopher-physicist Mach, a conflict emerged which has now proved almost meaningless. Ostwald was supported by a number of German chemists, Mach by a few physicists, but the majority of scientists supported the kinetic theory.

Clark [6], who himself supported the thermodynamic school, quotes a number of commentators:

> This opposition to the kinetic theory is often attributed by modern commentators to the existence of a fashionable philosophy of science which dismissed as unacceptable any theory based on speculative unobservable entities, such as atoms. For instance Einstein remarked of Ostwald and Mach that: 'The prejudices of these scientists against the atomic theory can be undoubtedly attributed to their positivistic philosophical views. This is an interesting example of how philosophical prejudices hinder a correct interpretation of facts even by scientists with bold thinking and subtle intuition.'

The same sentiment is echoed by Brush when he writes:

> Those scientists who did suggest that the kinetic theory be abandoned in the later 19th century did so not because of empirical difficulties but because of a more deep seated purely philosophical objection. For those who believed in a positivist

> methodology, any theory based on invisible and undetectable atoms was unacceptable.

Similarly Jaynes argues that [6]:

> ... the rise of the school of energetics championed by Mach and Ostwald, represents an early attempt of the positivist philosophy to limit the scope of science. This school held that to use modern terminology, the atom was not an observable, and that physical theories should not, therefore, make use of the concept.
>
> This thesis of philosophical prejudice is introduced to explain away objections to the kinetic theory, in terms of an extra-scientific, external influence, namely the dominance of an anti-atomistic, sensationalist philosophy. Other external explanations have been employed to account for the opposition to kinetic theories at the turn of the century. For example D'Abro writes of those who preferred thermodynamics to kinetic theories:
>
> ... their criticisms seem to have been dictated by preference rather than by reason. At all events their views found favour with many physicists who, though of first-rate ability in the experimental field, never evidenced any particular liking for mathematics ... one cannot help suspecting that the more difficult mathematical techniques of the microscopic theory may have influenced their philosophy.

It is noticeable that more chemists than physicists adopted the thermodynamic side in this controversy, particularly in Germany. This can easily be explained in terms of the aims of the two sciences. Chemistry is involved with macroscopic phenomena. Physics seeks the ultimate explanations of scientific entities. Clark, however, does not accept the general view [7]:

> I shall argue on the contrary, that the history of what is now called the kinetic theory must be appraised as the development of a powerful research programme, which after some early notable successes was degenerating in the last decade of the nineteenth century. In contrast the research programme of pure thermodynamics was progressive from its inception.

Basically the problems of kinetic theory reduced to that of the ratio of specific heats. Let us repeat the problem. A monatomic gas should have a ratio $C_p/C_v = 1.667$, with three degrees of freedom. A diatomic gas and a polyatomic gas should have three translational, three rotational and possibly some vibrational degrees of freedom. These would lead to a specific heat ratio of 1.333 or less.

Experimentally the ratio for a diatomic gas came out at about 1.4 for oxygen, nitrogen, NO and chlorine. This implied that only five degrees of freedom were operative, instead of the six or more expected. The discrepancy was taken very seriously even by supporters of kinetic theory. In 1900, Kelvin, for instance, referred to it as one of the two 'great clouds' hanging over physics (see the appendix at the end of this chapter) although quantum theory had already been discovered. Rayleigh wrote similarly at about the same time.

Maxwell himself, much earlier, in 1877, wrote of Boltzmann's attempt to eliminate the third rotational degree of freedom and the vibrational degrees by postulating perfect elasticity in the junction between the atoms of a perfectly elastic gas: 'The atoms of diatomic gases are elastic, but they cannot be perfectly elastic, since the internal atoms must vibrate on collision to give spectral lines' [8]. Maxwell was wrong here, since that is not the process by which spectral lines are emitted, but the source of his error was not known at the time. Kelvin, Loschmidt and Poincaré also rejected Boltzmann's theory on dynamical grounds.

So Boltzmann was under attack, not only from the followers of Ostwald and Mach, but from his own friends and colleagues. He himself valued the support of Max Planck, in the period 1880–1900, more than anyone else, but Planck was going through a difficult phase. In 1891 he said: 'Despite a short meteoric rise in the early sixties, every attempt at elaborating the theory has not only not led to new physical results but has run into overwhelming difficulties' [9].

Were these difficulties really overwhelming? There was a problem with the second law of thermodynamics but basically the overwhelming difficulty was the ratio of specific heats. It does not seem quite that serious. For forty years we have had the problem of the infinities in quantum electrodynamics, but this has not interfered with the award of Nobel Prizes for the theory. The locality problem in quantum theory has been with us for sixty years but is still not solved. The outstanding problems in contemporary cosmology and elementary particles are too many to enumerate, yet the theories are in current use.

So why was the specific heat problem regarded as so crucial? Was it a consequence of the arrogance which had crept into physics at the end of the nineteenth century, the belief, as Rayleigh put it, that the only future for physics was the more and more accurate determination of physical quantities. There does seem to be a tendency as each century creeps to its close to announce that the end of physics is nigh. Such sounds have been heard recently.

However, while such statements may have been characteristic of Rayleigh, and, to a lesser degree, Kelvin, one would not expect them of Maxwell. He was not only gifted with prodigious insight, but also had the virtue of scientific humility to a great degree. In this respect he resembled Kepler, rather than Galileo.

The answer to the specific heat problem is not difficult, though it had to wait until the clarifications of Bohr to emerge clearly. Flowers and

Mendoza put it well [10]:

> To explain the unexpectedly low specific heats of gases at ordinary temperatures we have to invoke the fact that the energy of a rotating body is quantised. It can only possess discrete values and can only change by discrete amounts. For monatomic helium molecules, the steps are of the order of 10 eV in size; thus the minimum energy of rotation that a helium atom may possess is of this order. If its total average energy is very much less than this amount it is highly improbable that it is rotating at all. Since kT at room temperature is of the order of 1/40 eV, it is evident that rotation of the atoms in helium at room temperature does not take place.

The probability of excitation depends on the factor $\exp(-h\nu/kT)$, which is simply related to the Boltzmann factor. A similar treatment is given for rotation. Application of quantum theory therefore, solves the anomaly. But this was not fully realized until about 1915, too late for Boltzmann.

Meanwhile, work had been proceeding on thermodynamics, notably by Planck, Ostwald and Gibbs. Planck's position in the period 1880–1900 was somewhat ambivalent. He was not a positivist and his approach was quite different from that of Ostwald, but he could not accept kinetic theory as it then stood. At the same time he used the Boltzmann factor in his quantum theory; though he did not entirely believe the outcome. Ostwald, according to Clark [11]:

> ... attempted in the research programme of energetics to derive the two laws of thermodynamics as consequences of the properties of the fundamental subject energy, the aim of the programme being to divest thermodynamics of its phenomenological character and thus to arrive at new laws, inaccessible by pure thermodynamic techniques.

Mach was also active at this time [12]:

> The basis of Mach's criticism of the kinetic programme was that programme's heuristic and empirical degeneration. His early defence of the programme, for example in 1863, relies on the range and number of new facts deduced by employing the theory. In the preface he wrote of the theory, 'one may accept the atomic theory ... as a formula which has already led to many results and will continue to do in the future'. Mach's opposition to atomism stems from its subsequent failure as a heuristic tool, e.g. in its persistent failure to deal with the specific heat problem. He came to regard it as a model for representing the facts, which would in accordance with the principle of economy of thought be subsequently replaced by a complete theory.

This may represent Mach's physical objections to the theory, but there was also the 'metaphysical' character of the atom, which as a good positivist he could not accept. He and Ostwald and their supporters appeared at conferences, and wrote papers and books supporting macroscopic thermodynamics and attacking kinetic theory.

Weinberg, who strongly dislikes positivism, has the following to say in his book *Dreams of a Final Theory* [13]:

> Positivism was at the heart of the opposition to the atomic theory at the turn of the twentieth century. The nineteenth century had seen a wonderful refinement of the old idea of Democritus and Leucippus that all matter is composed of atoms, and the atomic theory had been used by John Dalton and Amadeo Avogadro and their successors to make sense of the rules of chemistry, the properties of gases and the nature of heat. Atomic theory had become part of the ordinary language of physics and chemistry. Yet the positivist followers of Mach regarded this as a departure from the proper procedure of science because these atoms could not be observed with any technique that was then imaginable. The positivists decreed that scientists should concern themselves with reporting the result of observation, as for instance that it takes 2 volumes of hydrogen to combine with 1 volume of oxygen to make water vapour, but they should not concern themselves with speculations about metaphysical ideas that this is because the water molecule consists of two atoms of hydrogen and one atom of oxygen, because they could not observe these atoms or molecules. Mach himself never made his peace with the existence of atoms...
>
> The resistance to atomism had a particularly unfortunate effect in retarding the acceptance of statistical mechanics, the reductionist theory that interprets heat in terms of the statistical distribution of the energies of the parts of any system. The development of this theory in the work of Maxwell, Boltzmann, Gibbs and others was one of the triumphs of nineteenth century science, and in rejecting it the positivists were making the worst sort of mistake a scientist can make: not recognising success when it happens.

Boltzmann, as we have seen, had already been deserted by Maxwell, Kelvin, Poincaré, Loschmidt, and, he felt, by Planck; though Planck was silent on the specific heat question. Ultimately Planck was to save kinetic theory, though no one realized it until Einstein's paper on specific heats of solids in 1908, where he showed that frozen degrees of freedom were due to quantum limitations and would become unfrozen with a rise in temperature. A similar effect occurred in gases. There was no real conflict between kinetic theory and thermodynamics, but it required quantum theory in Einstein's interpretation to show it.

Boltzmann's grandson, Dieter Flamm, writing in the Schrödinger centenary volume [14] shows that the controversy did not end with Boltzmann's death, but now the protagonists were Mach and Planck. Ostwald was converted to the existence of atoms by the success of Einstein's work on Brownian motion in 1908. Planck too was totally converted to the kinetic theory by this time, and launched a violent attack on the positivists in 1908. Flamm says:

> Altogether, Vienna in 1911 when Schrödinger started his career as a scientist was a stronghold of the followers of Boltzmann ... evidence for the atomic structure of matter had been gathered from the study of radioactivity and of fluctuations [Brownian motion] ... but Mach's reputation was high as is witnessed by the fact that he had been nominated for the Nobel Prize...

Planck's paper launched a new controversy. In 1912 Mach's followers founded an association for positivistic philosophy, with a journal, but by now the controversy was dead as far as physics was concerned.

5.4. Conduction in gases

Meanwhile a totally different field was emerging in another branch of physics, the study of the conduction of electricity in gases. This complicated and messy subject dates from the 1870s, and no one looking at it would expect to get sensible results from it. It produced, between 1896 and 1900, x-rays, the electron, in a sense the proton and the beginnings of mass spectroscopy. Weinberg says [15]:

> Positivism did harm in other ways that are less well known. There is a famous experiment performed in 1897 by J J Thomson, which is generally regarded as the discovery of the electron.

He then describes the Thomson experiment:

> Yet the same experiment was done in Berlin at just about the same time by Walter Kaufmann. The main difference between Kaufmann's experiment and Thomson's was that Kaufmann's was better. It yielded a result for the ratio of the electron's charge and mass that today we know was more accurate than Thomson's. Yet Kaufmann is never listed as a discoverer of the electron, simply because he did not think that he had discovered a new particle. Thomson was working in an English tradition going back to Newton, Dalton and Prout—a tradition of speculation about atoms and their constituents. But Kaufmann was a positivist; he did not believe that it was the business of physicists to speculate about things that they could not observe. So Kaufmann did not report that he had discovered a new kind of particle, but only

that whatever it is that is flowing in a cathode ray tube, it carries a certain ratio of charge to mass.

At the same time, the similarly messy subject of radioactivity, in conjunction with long-standing spectroscopic observations, opened up the structure of the atom. Rutherford was predominant in the field. After elucidating the chemistry of the radioactive series, for which he was awarded a Nobel Prize in Chemistry (he never received one in physics), he embarked on the search for the structure of the atom. J J Thomson had put forward a model in which electrons were embedded in a mysterious, positively charged sphere of fluid. The sphere was about the size of the atoms which we have been discussing, a few ångströms in diameter. Rutherford realized that by firing alpha particles at a thin layer of a material, he could, by observing the scattering in traversal, deduce some information about atomic structure. The angular distribution of Thomson's atom should produce a narrow Gaussian curve, and most particles followed this law, but a few were scattered over much greater angles, and even bounced back off the foils. The only explanation was a highly concentrated nucleus of charge only one ten-thousandth of the diameter of the atom. Rutherford constructed a theory which succeeded in predicting the dependence on the element of which the scattering foil was made and the distribution of scattering angles. The atom was nearly all empty space, but to explain the ordinary behaviour of atoms that space must be filled. This was done by Niels Bohr. By his planetary model, in which electrons revolve around a nucleus of protons, Bohr was able to explain the spectrum of hydrogen and hydrogen-like atoms such as singly ionized helium. But why did the electrons not spiral into the nucleus as demanded by classical electrodynamics? Bohr cheerfully ignored this problem, which was eventually solved in the 1920s using quantum mechanics.

This brought to an end the atomic controversy at least in its classical form. More problems were to come, arising from the work of Einstein, Podolsky and Rosen in the 1930s, and a new positivism was born, but Boltzmann, driven to suicide by his critics in 1906, was vindicated, and his statistical mechanics became a cornerstone of physics. A synthesis of his ideas was achieved in statistical thermodynamics.

5.5. Conclusion

If the history of kinetic theory were a guide it might seem better if no philosophy were applied to physics at all, but in fact philosophy is unavoidable. Science creates philosophy whether or not the scientist wants it. Any judgment about the result of an experiment involves philosophy, though perhaps in a very simple form. The assent given to the proposition that the voltmeter is reading 50 V when that is how the eye synthesizes it and the nervous system processes it, is an act of faith with philosophical content.

More dangerous, as we have seen, is the application of a philosophical system such as positivism to the conclusions of a branch of physics. Yet a kind of positivism is essential in quantum mechanics. We talk about entities which are in principle observable, but which we cannot hope to observe directly with the naked eye. In spite of this, most experimental physicists tend to adopt a practical straightforward realism, and leave philosophical problems to elderly theorists. Very few physicists impose a taboo on the derivation of, say, atomic properties from observation of macroscopic quantities. If our experimental and theoretical techniques are analysed precisely, they tend to turn out to be somewhat eclectic. We speak of quantities which are observable and unobservable in the same breath. Only when challenged, as by Einstein, Podolsky and Rosen, do we try to analyse what precisely we are doing. In fact the assertion of the atomic hypothesis is itself the affirmation of a philosophical position in which atoms are possible.

Appendix: Kelvin's view of the problem

On 27 April 1900, Lord Kelvin gave a famous lecture at the Royal Institution in London. Its title was: 'Nineteenth century clouds over the dynamical theory of heat and light'. The first part was an acute analysis of the problems raised by the Michelson–Morley experiment on the propagation of light. The second part dealt with kinetic theory and particularly the ratio of specific heats. An extract from the lecture is given below [16].

23. Now for the application of the Boltzmann–Maxwell doctrine to the kinetic theory of gases: consider first a homogeneous single gas, that is, a vast assemblage of similar clusters of atoms moving and colliding as described in the last sentence of 19; the assemblage being so sparse that the time during which each cluster is in collision is very short in comparison with the time during which it is unacted on by other clusters, and its centre of inertia therefore moves uniformly in a straight line. If there are i atoms in each cluster, it has $3i$ freedoms to move, that is to say, freedoms in three rectangular directions for each atom. The Boltzmann–Maxwell doctrine asserts that the mean kinetic energies of these $3i$ motions are all equal, whatever be the mutual forces between the atoms. From this, when the durations of the collisions are not included in the time-averages, it is easy to prove algebraically (with exceptions noted below) that the time-average of the kinetic energy of the component velocity of the inertial centre, in any direction, is equal to any one of the $3i$ mean kinetic energies asserted to be equal to one another in the preceding statement. There are exceptions to the algebraic proof; but nevertheless the general Boltzmann–Maxwell doctrine includes the proposition, even in those cases in which it is not deducible algebraically

from the equality of the $3i$ energies. Thus, without exception, the average kinetic energy of any component of the motion of the inertial centre is, according to the Boltzmann–Maxwell doctrine, equal to $1/3i$ of the whole average kinetic energy of the system. This makes the total average kinetic energy, potential and kinetic, of the whole motion of the system, translational and relative, to be $3i(1+P)$ times the mean kinetic energy of one component of the motion of the inertial centre, where P denotes the ratio of the mean potential energy of the relative displacements of the parts to the mean kinetic energy of the whole system. Now, according to Clausius' splendid and easily proved theorem regarding the partition of energy in the kinetic theory of gases, the ratio of the difference of the two thermal capacities to the constant-volume thermal capacity is equal to the ratio of twice a single component of the translational energy to the total energy. Hence, if according to our usual notation we denote the ratio of the thermal capacity pressure-constant to the thermal capacity volume-constant by k, we have

$$k-1 = 2/[3i(1+P)].$$

[In modern terms, what Kelvin is saying is that if R is the gas constant, $C_p = C_v + R$. Then: $C_p/C_v = 1 + R/C_v = 1 + 2/n$ where n is the number of degrees of freedom associated with C_v. This is so because R has energy equivalent to two degrees of freedom.]

24. *Example 1*. For the first and simplest example, consider a monatomic gas. We have $i = 1$, and according to our supposition (the supposition generally, perhaps universally made) regarding atoms, we have $P = 0$. Hence $k-1 = 2/3$.

This is merely a fundamental theorem in the kinetic theory of gases for the case of no rotational or vibrational energy of the molecule; in which there is no scope either for Clausius' theorem or for the Boltzmann–Maxwell doctrine. It is beautifully illustrated by mercury vapour, a monatomic gas according to chemists, for which many years ago Kundt, in an admirably designed experiment, found $k-1$ to be very approximately $2/3$: and by the newly discovered gases argon, helium, and krypton, for which also $k-1$ has been found to have approximately the same value, by Rayleigh and Ramsay. But each of these four gases has a large number of spectrum lines, and therefore a large number of vibrational freedoms, and therefore, if the Boltzmann–Maxwell doctrine were true, $k-1$ would have some exceedingly small value, such as that shown in the ideal example of 26 below. On the other hand, Clausius' *theorem* presents no difficulty; it merely asserts that $k-1$ is necessarily less than $2/3$ in each of these four cases, as in every case in which there is any rotational or vibrational energy whatever; and proves, from the values found experimentally for $k-1$ in the four gases, that in each case the total of rotational and vibrational energy is exceedingly small in comparison with the translational energy. It justifies admirably the chemical doctrine that mercury vapour is practically a monatomic gas, and it proves that argon, helium, and krypton are also practically monatomic, though none of these gases has hitherto shown any chemical affinity or action of any kind from which chemists could draw any such conclusion.

But Clausius' theorem, taken in conclusion with Stokes' and Kirchoff's dynamics of spectrum analysis, throws a new light on what we are now calling a 'practically monatomic gas'. It shows that, unless we admit that the atoms can be set into rotation or vibration by mutual collisions (a most unacceptable hypothesis), each atom must have satellites connected with it (or ether condensed into it or around it) and kept, by the collisions, in motion relatively to it with total energy exceedingly small in comparison with the translational energy of the whole system of atom and satellites. The satellites must in all probability be of exceedingly small mass in comparison with that of the chief atom. Can they be the 'ions' by which J J Thomson explains the electric conductivity induced in air and other gases by ultra-violet light, Rontgen rays and Becquerel rays?

Finally it is interesting to remark that all the values of $k-1$ found by Rayleigh and Ramsay are somewhat less than 2/3; argon 0.64, 0.61; helium 0.652; krypton 0.666. If the deviation from 0.667 were accidental they would probably be some in defect and some in excess.

Example 2. As a next simplest example let $i = 2$, and as a very simplest case let the two atoms be in stable equilibrium when concentric, and be infinitely nearly concentric when the clusters move about, constituting a homogeneous gas. This supposition makes $P = 1/2$, because the average potential energy is equal to the average kinetic energy in simple harmonic vibrations; and in our present case half the whole kinetic energy, according to the Boltzmann–Maxwell doctrine, is vibrational, the other half being translational. We find $k - 1 = 2/9 = 0.2222$.

Example 3. Let $i = 2$; let there be stable equilibrium, with the centres C, C′ of the two atoms at a finite distance a asunder, and let the atoms be always nearly at this distance asunder when the atoms are not in collision. The relative motions of the two atoms will be according to three freedoms, one vibrational, consisting of very small shortenings and lengthenings of the distance CC′, and two rotational, consisting of rotations around one or other of two lines perpendicular to each other and perpendicular to CC′ through the inertial centre. With these conditions and limitations, and with the supposition that half the average kinetic energy of the rotation is comparable with the average kinetic energy of the vibrations, or exactly equal to it as according to the Boltzmann–Maxwell doctrine, it is easily proved that in rotation the excess of CC′ above the equilibrium distance a, due to centrifugal force, must be exceedingly small in comparison with the maximum value of $CC' - a$ due to the vibration. Hence the average potential energy of the rotation is negligible in comparison with the potential energy of the vibration. Hence, of the three freedoms for relative motion there is only one contributory to P, and therefore we have $P = 1/6$. Thus we find: $k - 1 = 2/7 = 0.2857$.

The best way of experimentally determining the ratio of the two thermal capacities for any gas is by comparison between the observed and the Newtonian velocities of sound. It has thus been ascertained that, at ordinary temperatures and pressures, $k - 1$ differs but little from 0.406 for common air, which is a mixture of the two gases oxygen and nitrogen, each diatomic according to modern chemical theory; and the greatest value that the Boltzmann–Maxwell doctrine can give is the 0.2857 of Example 3. This notable discrepancy from observation suffices to absolutely disprove the Boltzmann–Maxwell doctrine...

[End of extract.]

Reading this today, it seems astonishing that a great physicist, even one seventy-six years old, could be so blind. What Kelvin is objecting to is not the postulate of atoms, as Mach was. Nor indeed is he objecting to kinetic theory. For him, the 'Boltzmann–Maxwell doctrine', is the equipartition of energy. In a later table he gives the value of $k - 1$ for four diatomic gases: three of them are very close to 0.40, yet this passes without comment.

One is tempted to repeat Weinberg's remark about the positivists—'making the worst kind of mistake a scientist can make: not recognising success when it happens'—Kelvin of course was not a positivist.

Chapter 6

N-RAYS

Generally physicists are fairly phlegmatic, but occasionally for a short time they rush like a horde of lemmings into the fjords of self-deception. The classical example of this was the observation of N-rays, though there were other contenders: the preoccupation of the Royal Society in the 1870s with the properties of hot ice, and the excitement over the lambda-meson in the 1940s. N-rays will be for ever associated with the name of Rene Prosper Blondlot.

Blondlot was a professor at the University of Nancy in north-eastern France, east of Paris and quite close to the German border. The city is industrial with a population of about a hundred thousand. When the story of N-rays begins, in about 1903, the University was a respected middle-grade institution, and Blondlot was a corresponding member of the Academy of Sciences, a coveted distinction. His scientific credentials were impeccable and solid, a physicist of the classical school. However, the times were not classical: physics, as we have seen elsewhere in this book, was in some disarray. In 1895, Lord Rayleigh, a great physicist, remarked that the future of physics lay in the fifth significant figure [1]. That is to say, the basic laws were all known, all that was necessary was to make more and more accurate measurements. Within six months the roof of the classical structure fell in: Röntgen had discovered x-rays, J J Thomson the electron, Henri Becquerel radioactivity, further refined by the Curies and in 1900 Planck produced the quantum theory.

Popular attention and that of most scientists was focused on x-rays, and many observations were being made. It is not surprising therefore that Blondlot turned first to x-rays for his equipment. He also used methods best suited to optical spectroscopy, to infrared and to radio; and his experimental system was a bewildering amalgam of all these.

A most important feature was the use of a darkened room and a dark-adapted eye for making measurements. The human eye is an astonishingly sensitive instrument. If it is sensitized by a burst of about ten photons, it is then capable for some time afterwards of detecting a

Figure 6.1. *Rene Prosper Blondlot.*

single photon. It must first of all be dark-adapted by resting in total darkness for about forty-five minutes, but though sensitive, the eye is erratic. Even in the dark it suffers from background signals, some of them genuine light, some noise in the processing nervous system. It was shown in 1960 that cosmic rays passing through the eye produce flashes of light which are detectable, and this will happen about ten times a minute at sea level. It cannot be screened out by ordinary buildings, and one would have to go hundreds of metres underground to reduce the cosmic-ray flux significantly. More serious is the noise in the retina and the processing cells behind the retina. These give rise to flashes of light and sometimes peculiar patterns, familiar to anyone who has ever closed their eyes in a dark room at night. As already mentioned, there is also a background of light produced by cosmic rays passing through the air in a room. For a ten-metre cube there will be about a million photons a minute produced in the air. This is a very small amount of light, but it will always be present.

In 1903, however, the human eye, with all its limitations, was the most sensitive detector of electromagnetic radiation in the visible part of the spectrum. Photography was not sensitive enough to compete, though Blondlot did do some experiments with photographic film.

The source of N-rays was an Auer lamp (Freiherr Auer von Welsbach was an aristocratic and ingenious Austrian chemist who invented a number of things including the incandescent gas mantle, still used in portable lamps, the alloy from which cigarette lighter flints are made and the Osmium lamp). The lamp was encased in an iron tube which had a small aperture. This aperture was closed by a window made of aluminium about one-eighth of an inch thick. In preliminary experiments it had been found that N-rays can pass through aluminium but not through iron (as it happens, x-rays are absorbed more strongly by iron than by aluminium).

So what were N-rays? This was never quite clear. It was made more difficult for Anglo-Saxons because virtually all of the papers were in French. However, the journal *Nature* had at that time the commendable custom of summarizing the contents of foreign journals and the proceedings of foregn Academies; so the volumes of *Nature* from 1903 to about 1905 contain the history of the subject. Thus in October 1903 it said [2]:

> Several papers on the so-called N-rays discussed by M Blondlot are printed in the *Journal de Physique* for August. M Blondlot shows that these rays are of common occurrence, being emitted by an Auer lamp and an incandescent silver lamina, and being present in sunlight. M G Sagnac describes determinations of the wavelength of these rays by means of their diffraction. It appears that the rays in question are about two octaves below the Rubens infra-red rays and intermediate between these and the Hertzian radiation of Lampa. The wavelength is about 0.2 of a millimetre.

The analysing device used was a prism spectrometer, but the prism was made of aluminium instead of glass or quartz. Apart from the aluminium prism the spectrometer resembled those used in laboratories for the study of visible spectra. N-rays cannot be seen so we have to borrow another technique from x-rays, and use a thread impregnated with calcium fluoride. This glows when the N-rays strike it. It is difficult to follow the thought processes which determine the choice of equipment. There is a surreal quality about the whole adventure. For instance it was discovered early on that sunlight contains N-rays. They could even be stored. If a brick were wrapped in black paper and placed in the sunlight it would become charged with N-rays. These could be detected by holding the brick to the head of an observer. He was then able to see more clearly in the dark. Now it is perfectly true that far-infrared radiation is a component of sunlight. Moreover if you put a brick wrapped in black paper in the Sun, it will get hotter. If you hold the brick against someone's head, it will heat up his head. It is conceivable that this might change the threshold sensitivity of the eye slightly, or increase the noise level, so claims of this kind are very difficult to disprove. Such features will recur as we go into more detailed studies of the experiments.

Practically all the positive effects were seen in France. In Germany, England and the United States no one could reproduce the effects and generally no one saw anything at all. Professor Rubens of Berlin, a distinguished expert on infrared radiation, was particularly unfortunate. He was summoned to Potsdam by the Kaiser, and reprimanded for failing to see the N-rays. Vague threats about his academic future were made. Elsewhere a graduate student who had been allocated the task of repeating one of Blondlot's experiments suffered agonies because of his inability to see anything. In despair he decided that he would never be a physicist, gave up, and went into engineering. He became very successful and made far more money than he would have made in physics, but always regretted, he said to this author, the twist of fate which had driven him, undeservedly, out.

The observed effects of N-rays are best described in the words of the people who witnessed them, as summarized by *Nature*:

> *Academy of Sciences, Paris, 2 November 1903. On new effects produced by the N-rays, generalisation of the phenomena originally observed. M R Blondlot.*
>
> The N-rays are rays given off by various sources of light, capable of passing through an aluminium screen, and recognisable by their action on a small electric spark or upon a feebly phosphorescent screen. It has now been found that these rays cause a slight but distinct increase in the luminosity of a feebly illuminated paper screen, and this effect is retained by the rays after reflection at a polished metallic surface [3].
>
> *Academy of Sciences, 9 November 1903. On the storage of the N-rays by certain bodies. M R Blondlot.*
>
> The rays from various sources of light, after being filtered through an aluminium screen, possess the property of increasing the luminosity of a feebly phosphorescent screen. It was noticed that when a quartz lens was used this effect continued after the source of light, an incandescent mantle, was extinguished, and it was then found that quartz, iceland spar, fluorspar and various other substances possessed the same property. The rays are stored throughout the same mass and take some time to penetrate [4].
>
> *Comptes Rendus, 24 December 1903.*
>
> M A Charpentier shows that these rays are also emitted by the human body especially by the muscles and nerves [5].
>
> *Academy of Sciences, 25 November 1903. On the reinforcement of the bundle of light rays upon the eye when accompanied by N-rays. M Blondlot.*
>
> It has now been found that if the N-rays are directed toward the eye a similar reinforcement is obtained [6].

Comptes Rendus, 24 December 1903.

M Blondlot has found that bodies in a state of strain such as tempered steel and unannealed glass give off these rays spontaneously and continuously at the ordinary temperatures [7].

Academy of Sciences, 24 February 1904.

A Charpentier has found that a body emitting the N-rays when brought near the nose sensibly increases the intensity of the sensation of smell [8].

Academy of Sciences, 14 March 1904. M Blondlot.

While the N-rays increase the N-rays emitted normally by the screen they diminish the light emission obliquely [9].

It was inevitable, perhaps, that having discovered N-rays, anti-N rays would also be discovered:

Academy of Sciences, 29 February 1904. M Blondlot.

On a new species of N-rays. These rays diminish the luminosity of a phosphorescent calcium screen, instead of increasing it. They are present, along with the N-rays, in the light of a Nernst lamp [10].

Academy of Sciences, 15 February 1904. A Charpentier.

A description of a series of experiments on the conduction of N-rays along copper and silver wires. A piece of string moistened with collodion containing calcium sulphide in suspension, also conducts like a metallic wire [11].

Academy of Sciences,. 21 January 1904. The emission of N-rays by plants. Eduard Meyer.

N-rays are emitted by plants, the emission being a function of the nutritive activity or evolution of the plant [12].

The following is perhaps the most sophisticated work published on N-rays:

Academy of Sciences, 14 March 1904. H Bagard.

The rotatory power of solutions of sugar, tartaric acid and turpentine for eight groups of N-rays has been examined, the rays being distinguished by the refractive indices through aluminium and by their wavelengths. For turpentine and sugar the rotations are in the same sense as with ordinary light, and are normal in that the rotation varies in the inverse sense with wavelength. For tartaric acid, which produces a dextro rotation with ordinary light, the rotation is to the left in N-rays. The rotations observed with the latter are several hundred times greater than with ordinary light; thus a solution giving +4°42′ in a column of 20 cm length with ordinary light gave −138° with N-rays in a column only 0.055 cm long. It was found that the rotations produced were proportional to the thickness of the solution within the limits of experimental error [13].

Academy of Sciences, 25 January 1904. M Lambert.

The N-rays are produced by the action of ferments, the effect being particularly marked for the digestive ferments of albumenoid materials [14].

Academy of Sciences, 21 April 1904. Julien Meyer.

The N′-rays discovered by Blondlot, the effects of which are the inverse of the N-rays, are given off by vacuous glass tubes and possess a greater penetrating power than the similar rays given off by a Nernst lamp. Certain substances appear to possess the power of storing up these rays, aluminium being a notable example, and then emit them for as long as twenty four hours later [15].

Academy of Sciences, 21 April 1904. A Colson.

By means of the effects on a phosphorescent screen the author has been able to detect differences in the interaction of solutions of potash and zinc sulphate according to the order in which they are mixed, and these differences have been subsequently borne out by their chemical behaviour [16].

Altogether there were thirty-five papers on N-rays noted in volume 69 of *Nature*. Those quoted here were all positive, but there were about the same number of papers which simply noted failure to detect anything. These were from German and British authors. The tide was to turn in volume 70 (May to October 1904), which had only ten entries. One of these is interesting:

Academy of Sciences, 16 May 1904. A Broca and A Zimmern.

From the preliminary observations the authors conclude that the examination of the spinal cord by means of the N-rays allows of the existence on the living man of the existence of medullary centres, and even to gain some idea of their degree of activity [17].

In 1904, Rubens and a number of other physicists contacted R W Wood, a distinguished American physicist who was then in Brussels.

Wood was multilingual and highly respected in Europe and America. He was born in 1868, in Concord, Massachusetts, and educated at Harvard, Chicago and Berlin. From 1901 to 1938 he was Professor of Physics in Johns Hopkins. He was working in Brussels on his famous book *Physical Optics* when he met Rubens and his colleagues. It was suggested that he should go to Nancy and investigate the N-rays. He was somewhat reluctant to do so, saying that Rubens had suffered most, perhaps he should go. Finally, however, it was decided that Wood should go since he had not published any results, positive or negative, on N-rays.

In Nancy he was greeted warmly by Blondlot, who was anxious to show his results and techniques, and seemed totally convinced. Blondlot was working with an assistant and Wood agreed with Blondlot that their conversation should be in German, a language that the assistant did not know. Nor did the assistant know that Wood spoke French. The demonstrations were carried out by Blondlot and his assistant in a darkened room. While the assistant was describing the N-rays, Wood surreptitiously removed the prism from the spectrometer. The assistant went on describing them, which of course he should not have been able to do. Wood then replaced the prism, but was spotted by the assistant, who said to Blondlot in French: 'He has disturbed the prism'. In fact Wood had replaced the prism correctly, but the assistant now claimed to be unable to see the N-rays. He should have done, since the system was back in its original state. Wood was also shown an experiment where N-rays were alleged to increase the illumination of a faint spark. He was not impressed. Also a steel file held close to a phosphorescent clock face was said to increase its visibility.

Wood wrote a letter to *Nature*, in which he did not give the names of the experimenters, and did not elaborate on the prism incident. He concludes his letter [18]:

> I am obliged to confess that I left the laboratory with a distinct feeling of depression, not only having failed to see a single experiment of a convincing nature, but with the almost certain conviction that all the changes in the luminosity or distinctness of sparks and phosphorescent screens (which furnish the only evidence of N-rays) are purely imaginary. It seems strange that after a year's work on the subject not a single experiment has been devised which can in any way convince a critical observer the rays exist at all.

There are some unexplained features. Blondlot showed an uncanny ability to set the spectrometer to the expected value, with an accuracy of 0.1 mm. There are a number of phenomena which might have simulated N-rays. Triboluminescence causes light from the breaking of glass and other materials, friction between two crystals, and even the breaking of dry wood. It is interesting that green wood did not emit N-rays. Electrostatic phenomena seen when adhesive tape is unstuck in the dark and the production of flashes of blue light when two sugar lumps are rubbed together are well known party tricks.

The imagination boggles, however, at the aluminium prism, and the use of sheets of wood for N-ray filters. Real signals may have been observed in some cases but they were not due to N-rays. After Wood's intervention the subject fell away rapidly, though Blondlot never lost his faith. His health declined, perhaps as a result of the debacle. The Academy of Sciences had agreed to give him a medal for N-rays. They did, but it was awarded for his life's work. Is there a moral to this story? It seems difficult to find. Other bizarre effects such as some radioactive phenomena were perfectly genuine. It is of course easy to laugh at the end of the twentieth century, with cooled phototubes and charge coupled device (CCD) infrared detectors freely available. Perhaps as a matter of *pietas*, some physicist using objective devices, should test some of the more plausible of the N-ray effects. Such effects may be due to heat. A schoolmaster in Hertfordshire, failing to find N-rays, observed that a very small intensity of heat increased the brightness of phosphorescent screens considerably. The Auer lamps must have generated a considerable amount of heat even when attenuated by the aluminium screens, but it seems very likely that most of the N-ray effects were imaginary. Let he that is without sin cast the first stone.

Chapter 7

EINSTEIN AND THE COPENHAGEN SCHOOL

Albert Einstein was born in Ulm on 14 March 1879. The town was on the banks of the Danube; a pleasant, rather sleepy place, largely untouched by industry. His parents were Jewish, but non-practising. The family had been in Swabia for many generations; but like many German-Jewish families of the time, had cousins all over Central Europe.

Unfortunately, when Albert was one, the family business failed and they had to move to Bavaria. They settled in Munich, the capital of an independent kingdom, but closely tied to the Greater Germany. It was strongly Roman Catholic, and when Albert went to school, he went to a Catholic school. He was not happy there, and it was there, at the age of five, that he first encountered anti-semitism. His father set up an electrical engineering company, which was very successful. Albert had a sister Maja who was two years younger, born in 1881. They were happy. Albert was not happy. Nor were his teachers happy with him. Most of his learning was done on his own at home. Mainly this was in mathematics, where he covered the high-school syllabus while still at primary school. But he was regarded as insubordinate, in a quiet way, and inattentive. He came within a short distance of expulsion. He did not play with his classmates; though later he was to form some friendships. Altogether, he was a most unsatisfactory child. Apart from mathematics, his talent was for the violin, which he learnt from the age of six. His younger sister Maja idolized him throughout her life.

At high school, the Luitpold Gymnasium, the pattern was repeated. In mathematics he was far ahead of his class, but in other subjects inattentive to an almost pathological degree. His parents had feared, up to the age of nine, that he might be handicapped. He reacted strongly against the Prussian style of highly disciplined teaching which was in vogue at the time. Whether modern child guidance would have recognized a genius of this kind is an interesting speculation. He read

Figure 7.1. *Albert Einstein (1879–1955).*

the philosophers as well as mathematics, whilst his father was trying to steer him towards a career in electrical engineering.

When he was fifteen a catastrophe struck. His father's business failed, and the family had to move to cousins in Milan. They left Albert behind to complete his education. After that he would face conscription into the German Army. He was in despair. One of his few close friends was Max Talmey, a medical student who had helped Albert since he was nine. Now graduated and gone away, Max's elder brother was a practising doctor in Munich. Albert persuaded him to prescribe a drier climate for his lungs and, more difficult, persuaded the school principal to give him a certificate of proficiency in mathematics. Armed with these, Albert set off for Switzerland.

He wanted to enter the famous ETH (Zurich Polytechnic), but failed the entrance examination. His examiner was so impressed with his mathematics, however, that he arranged for private tuition. At Albert's second try he passed easily and entered a physics course. It seems that he did not contemplate a full time career in research, but hoped for a more modest goal: teaching, with some time for research. This was of course the life of a university professor, so he was aiming quite high.

He lived a somewhat solitary life, but had some friends whom he would occasionally join for a drink at the Cafe Metropole, a favourite student refuge. His closest friends were the Swiss Marcel Grossman and a Serbian girl, Mileva Maric. Both were mathematicians. Since Einstein did not attend lectures regularly, his two friends had to coach him at weekends. Meanwhile he was wandering through the original sources, a thing rarely done by undergraduates. He was particularly fascinated by the positivist Ernst Mach. His attitude to Mach was ambivalent [1]. He responded to the positivist demand for experiment, but on the other hand he believed in atoms, which, as we have seen, Mach did not. The position of Mach's principle in general relativity is still controversial.

By the time Einstein graduated in 1900, he and Mileva were deeply in love, but they could not marry until Albert found a job. Mileva's parents were wealthy, but Albert and Mileva refused to live off them. Another difficulty arose. Something happened which Mileva described as 'intensely personal', but refused to discuss any further. It has recently been found that she had a child by Einstein in 1900. Albert found temporary jobs, first at the Zurich Observatory, then at a boy's school. He published his first paper, on capillarity. At about that time he became a Swiss citizen, but was rejected by the Army on medical grounds when he was called up for National Service.

Now he needed a job. In 1902, he was taken on by the Swiss Patent Office, as an examiner. He liked the work, and became quite good at it. Shortly after, he and Mileva were married. They had a son, Hans Albert, in 1904.

Now, Einstein was starting his own original work. In 1905, the '*Annus Mirabilis*', he produced three papers. One was on the photoelectric effect, and showed how Max Planck's quantum theory of the black body radiation of 1900 could be explained in terms of photons, particles of light. Since the wave theory of light was very well established, this meant that light had to be explained as a dual phenomenon, wave and particle. It was the origin of quantum mechanics. One of the most splendid things about this theory is that schoolchildren can understand it, yet the Planck theory was misunderstood by physicists for four years.

The next paper was on Brownian motion. This was an apparently trivial effect that had been known for eighty years without being understood. Small particles suspended in water or other fluids showed under the microscope a random continuous motion. Einstein showed that from this motion, quite accurate estimates could be made of the sizes of molecules. Only very approximate measurements had been available hitherto. This also fitted in with the much grander system of molecular physics proposed by Ludwig Boltzmann, whom we have already considered, and was evidence for the existence of atoms; which even then was not universally accepted.

However, the most original and far reaching of these remarkable papers was that on the electrodynamics of moving objects. If it had been submitted to a modern journal it would probably have been turned down, because it had no references and virtually no numbers or calculations. The editor of *Annalen der Physik* was more perceptive and published it. It goes back to the negative result of the Michelson–

Morley experiment of 1886, carried out in Cleveland, though it does not deal with it in detail.

Ever since the work of such people as Young, Fresnel and Lloyd in the early 1800s, which demonstrated the wave nature of light, it was assumed that the waves must propagate in a medium. They called this the ether. The ether had remarkable properties. It was so tenuous that solid bodies could move through it without restraint, yet to explain the very high velocity of light it had to be extremely rigid. Michelson and Morley, in a very sensitive experiment using the interference of two rays of light at right angles, looked for the speed of the Earth through the ether. They found no sign of the ether at all. The Irish physicist Fitzgerald suggested that one arm of the interferometer contracted slightly to counteract the expected shift. This gave the correct answer but was not explained by a theory. The same answer came in a logical way out of Einstein's theory, which was not very complicated.

In spite of its simplicity the special theory of relativity (that in which velocities are constant), remained controversial and when he received the Nobel Prize in 1922, it was for the photoelectric effect, not for relativity. In the meantime he had produced the general theory, which deals with accelerated bodies, and is far more difficult to understand. The delay was largely due to the conservatism of classical physicists, but undoubtedly there was an element of German anti-semitism involved.

Anti-semitism had been strong in Germany for many years; in the universities as elsewhere. People were passed over for promotion and failed to get professorships simply because they were Jewish. It was particularly virulent in the years immediately following the First World War. Nazism did not create it, therefore, but rather latched on to a thing that was already widespread in German society. Inflation made it worse.

Thus when Einstein leapt to world fame in 1919, there was some resentment in Germany. In Heidelberg, the most ancient of the German universities, Professor Lenard held sway. Lenard, a Nobel Prizewinner, was a right-wing nationalist, and was to figure prominently in the persecution of Professor Gumbel, a pacifist, and later in the campaign against Einstein.

The cause of Einstein's sudden popularity was the result of the British eclipse expedition of 1919. The general theory of relativity had in 1915 predicted that a light ray would be deflected in passing very close to a massive object such as the Sun. This was not new; Newtonian theory also predicted a deflection, but the relativistic one was twice the Newtonian one. Sir Arthur Eddington's expedition found a value in agreement with the relativistic prediction, and Einstein became a world figure overnight.

There followed a hectic ten years in round-the-world travel, lecturing to enormous audiences on relativity, as well as supporting pacifism and Zionism. His opinions were sought on every conceivable topic and his wit and good humour were tried to the limit.

But, as a distinguished Swedish physicist was once heard to remark, there were dark forces working against him in the Waterfalls Board. In Einstein's case, it was not the Waterfalls Board, but the German physics establishment. The prejudices were based, probably, most of all on jealousy, secondly on anti-semitism and, to a lesser extent, on a classical view of physics. They recall the verdict of the commission set up to investigate the case of Galileo: Copernicanism was 'philosophically ridiculous, and contrary to holy scripture'. In Einstein's case, the establishment thought relativity was 'philosophically ridiculous' and in its view contrary to holy scripture, equated with relativity being contrary to the racial theories cobbled together by Alfred Rosenberg. Meanwhile in the late 1920s and early 1930s, Einstein was riding high on world-wide acclaim. When the Nazis came to power in 1933, he was abroad, and no longer welcome at home in Germany. He spent the rest of his life in exile, settling at the Institute for Advanced Study in Princeton, New Jersey. After the Second World War he was offered the Presidency of Israel, but refused, preferring to stay away from post-war politics.

He had two main preoccupations in his later years: the unified field theory, and the difficulties of quantum mechanics.

The search for a unified field theory had been going on for a long time, some would say since the time of Archimedes. In the late 1920s and early 1930s, however, it had resolved itself as follows.

There are four kinds of force: gravity, the electromagnetic force, the weak nuclear force which causes beta decay, and the strong nuclear force. Einstein's object, which would mean in a sense the ultimate goal of physics, was to unify all of these in a single set of equations. He was not alone in this: Schrödinger in Dublin and many others were working towards the same goals. Most of the *'alte Herren'* of physics turned to this in their later days. The problem has still not been solved, though much progress has been made: Salam, Glashow and Weinberg received the Nobel Prize for their unification of the weak and electromagnetic forces in 1979. The strong force so far has partially resisted unification; and gravity, the weakest but most far-reaching force, still stands alone. Perhaps Professor Stephen Hawking's quantum theory of gravity is an important start. We do not yet know.

What is certain is that Einstein himself did not succeed. We know that he spent much of his time fighting the Nazis in every way that was possible for an elderly world figure: campaigns, speeches, money-raising, helping refugees. What occupied the rest of his time was the unified field theory, and the problems of quantum mechanics. The controversy about the latter was the second great conflict of his life, but it was carried on without the savage brutality of the Nazis. It was a civilized conversation between gentlemen. The main protagonist on the other side was Professor Niels Bohr of Copenhagen, and we have already reviewed part of the problem.

Quantum mechanics derived ultimately from the 1901 theory of Max Planck, and the demonstration by Einstein himself in 1905 that the photoelectric effect could only be understood in terms of quantum theory. There were other important steps: the discovery of the Compton effect in 1921 showed that an electromagnetic wave must be regarded both as wave-like and particle-like. Then Prince Louis de Broglie demonstrated that a particle of a given momentum must have a definite wavelength, which was confirmed by experiment.

In 1925–6, 'the deluge' threatened to sweep away the whole of classical physics. It could not, of course; momentum, energy, force and so on will withstand the most severe battering. But many of the older generation, who frankly did not entirely understand the mathematics of the new physics, feared the end was nigh, if not already come. Relativity, of course, had already raised its head, to the fury of Professor Lenard and his friends; but this new threat was more fundamental still. The whole of matter was under attack. One wonders what Ernst Mach would have thought; would he have been pleased? Some positivists certainly were, because of the new theory's insistence on observables. But there was more to it than that.

There were three strands to quantum mechanics. The first was the wave mechanics of Schrödinger and de Broglie, which is now thought of as being somewhat classical although it is the most commonly used model. In it, real waves resonate and interfere and generally do what classical light waves do. It is not clear what medium these waves are in. Sir Arthur Eddington used the forbidden word 'ether', but only, we assume, in fun. It is not in fact necessary to specify the medium. There is a wave equation and that is enough. Schrödinger's formulation is the most frequently used in day-to-day physics, but is not considered the most fundamental. Heisenberg's formulation is considered mathematically equivalent to the Schrödinger approach, but is more abstract and difficult. Finally there is Dirac's formulation, which is the most abstract but also the most powerful of all.

7.1. Niels Bohr

Niels Hendrik David Bohr was born in 1885, the son of Christian Bohr and Ellen Adler Bohr. His father was a professor of physiology in the University of Copenhagen. Niels and his brother both distinguished themselves athletically. His brother represented Denmark in the Olympic Games. Niels had a brilliant undergraduate career in science, took his Master's degree in 1909 and his PhD in 1911. His first published work was on vibrations in liquid jets. In 1912, the Carlsberg Foundation gave him a grant to study with J J Thomson at Cambridge. He found little going on and after a while left and joined Rutherford in Manchester, at that time the centre of world physics. He was given first a research post, then, after an absence in Denmark, a readership. This was a

Figure 7.2. *Niels Bohr (1885–1962).*

senior post for his age, and he was given a free hand by Rutherford. Most of the work in Manchester at that time was experimental or closely derived from experiment, so Bohr, a theoretician, had to look for problems himself. He chose the Balmer series. This, discovered by a Swiss schoolmaster in the 1880s, was a formula which fitted very closely the wavelengths of the lines in the hydrogen spectrum. Strikingly, it involved a combination of whole numbers, implying some fundamental law. Bohr applied quantum theory, and succeeded in explaining the formula, though with some strange consequences. We will consider this in more detail later. Returning to Denmark, he was appointed Professor of Theoretical Physics in 1916, and Director of the Institute for Theoretical Physics in 1920. The Institute became a world centre, and over the years large numbers of young scientists flocked to Copenhagen to work there. Bohr was a humane and friendly person, particularly with young physicists, who admired and respected him enormously. Einstein and Bohr had a deep mutual affection and respect for each other, in spite of their disagreements. They did not actually meet until 1920, but followed each other's work closely before that. Indeed the first disagreement went back nearly ten years before their first meeting. Bohr's main defect, though some considered it a virtue, was his extraordinary meticulousness. He was the despair of editors as he checked and rechecked up to nine sets of proofs of a journal article. This extreme care showed itself in what was known as 'Bohrspeak', a closely knitted language in which every word counted, and which required very careful listening.

In spite of his eminence and the many honours bestowed on him, he never lost his sense of wonder at the world. Einstein wrote of him after their first meeting: 'He seems to me like a sensitive child, who moves through the world in a trance'. His spoken English was very difficult to follow, so much so that when he gave the introductory address at the first Atoms for Peace Conference in 1956, the organizers laid on a simultaneous translation into English; although he was speaking in English.

7.2. Einstein and the old quantum theory

Any one of his 1905 papers would have satisfied Einstein's claim to greatness. Here we can follow only one of them, that on the photoelectric effect and the quantum theory. It is presented in elementary texts as if it had been immediately accepted by the physics community. In fact it took nearly twenty years, and the word 'photon' was not invented until 1926. Another false impression often given is that this was Einstein's only contribution to quantum theory. This also is far from the truth. He was a major author in quantum physics from 1905 to 1925 apart from his contributions to relativity and other topics.

The 1905 paper deals with the photoelectric effect, which had been studied carefully by Lenard in 1902. (Ironically, in the 1920s Lenard, as a leading Nazi professor at Heidelberg, was to become one of Einstein's most bitter enemies.) Lenard had found that the energy of the photoelectrons produced when light fell on a metallic surface was independent of the intensity but depended on the colour of the light. It was also found that there was no delay between the onset of the light and the emission of the electrons, even for very faint illumination. In classical wave theory these facts were inexplicable. It seemed that one electron in the metal was gathering all the energy in the light over a large area around it to acquire enough energy to knock it out of the surface. Why was one electron so privileged over millions of others? If we introduce the 'light-quantum' $h\nu$, however, all these problems disappear. The electron emitted is just the one hit by the quantum. The energy of the electron is determined by the energy of the quantum less the workfunction of the metal: the colour of the light is important because it determines the energy of the quantum.

In spite of these advantages, the light-quantum was not accepted. As late as 1913, Planck, Nernst, Rubens and Warburg proposed Einstein for membership of the Prussian Academy. After praising his achievements they said: 'That he may sometimes have missed the target in his speculations, as, for example, in his hypothesis of the light-quanta, cannot really be held against him, for it is not possible to introduce really new ideas even in the most exact sciences without sometimes taking a risk' [2].

On the other hand, Einstein's second venture into quantum theory, the specific heat of gases, was an immediate success. The problem went back nearly a hundred years. In 1819, Dulong and Petit measured the specific heats of a number of solids. They found that the value of C_V for most of the materials tried was approximately the same, at about 6 cal mol^{-1} °C^{-1}. In 1840 two Swiss physicists measured diamond and found a value of only 1.4 cal mol^{-1} °C^{-1}. Regnault found 1.8 over a slightly different temperature range, and Weber pointed out that if C_V varied with temperature, the two measurements could be brought into agreement. He set out to measure C_V over a wide temperature range, from 0 to 200 °C, but his measurements were held up because the spring came and there was no more snow for his ice calorimeter. He demonstrated, however, that there was a rapid change of C_V with temperature for diamond. In 1875 he extended his measurements to cover the temperature range −100 °C to +1000 °C, and found that C_V for diamond varied by a factor of 15 between these limits. At low temperatures, C_V is small, but at high temperatures it tends to the Dulong and Petit value. Dewar came to the same conclusion independently and in later work measured C_V for diamond at very low temperatures, from 20 to 85 K. He found values as low as 0.05 cal mol^{-1} °C^{-1}.

Boltzmann attempted to explain the Dulong–Petit law theoretically in 1866. After some difficulties had been surmounted, he was successful. In a solid there are no translational or rotational degrees of freedom, but vibration occurs. Each orthogonal direction has two degrees of freedom, one for potential and one for kinetic energy. Thus there are six degrees of freedom altogether, each with half the gas constant R; making $3R$, about 6 cal mol^{-1} °C^{-1}, as Dulong and Petit found. He was unable, however, to explain the low values of C_V. As with gases it appeared that some degrees of freedom were frozen or partially frozen.

Work continued during the nineteenth century, and by 1900 there was a solid body of experimental data. Diamond was the most extreme example of the variation, but in general it was found that light, hard materials showed low values of C_V, whereas heavy, soft materials like lead had high values at room temperature. The problem now was to explain this on an atomic scale. All materials tended to the Dulong–Petit value at high temperatures. At very high temperatures (over 1000 °C) values greater than the Dulong–Petit value are observed, but this was not well known when Einstein commenced his work on the subject in 1906. They were explained later in terms of the electrons in the solids and defects in crystals.

Basically Einstein adopted the Planck quantum formula and assigned a characteristic frequency of vibration to the material. With the limited degrees of freedom available he obtained a variation of specific heat similar to that found experimentally as the temperature rose from absolute zero. The agreement was not perfect and the Dutch physicist Peter Debye improved the theory by introducing a spectrum of frequencies. This gave almost perfect agreement with experiment. The Einstein theory was almost universally accepted especially as it did not make explicit use of the light-quantum.

In 1913, Einstein took up the problem which had bedevilled nineteenth-century physics and caused Boltzmann's suicide, the specific heat ratio for gases. He showed that the quantum expressions for the probability of excitation of a degree of freedom explained why some degrees of freedom were frozen. He was also aware of Rutherford's nuclear atom, which helped to explain which degrees of freedom would be frozen. In particular, for a diatomic molecule the line joining the nuclei would not be available for variations in rotation. The theory was successful. A more sophisticated treatment was promised but never actually appeared. In this year Einstein was appointed to a position in Berlin, moving there in 1914 from Prague.

The years 1915 to 1917 were largely taken up with general relativity, but in 1916 he worked on spontaneous and induced emission, work which many years afterward led to the laser. In 1917 Einstein became ill and did not fully recover until 1920. By that time he was a world celebrity, with the success of the British eclipse expeditions, which found values in agreement with relativity theory. Meanwhile Millikan had done some very precise work on the photoelectric effect, finding complete agreement with the theory of 1905. It did not resolve the mystery of the light quantum; even Einstein was unhappy with it.

In 1920, as we have said, Einstein finally met Bohr. They were greatly taken with each other. Bohr's work on the spectrum of hydrogen-like spectra had this in common with the theory of the photoelectric effect: both theories worked very well but no one understood what they meant, and they both involved, in a mysterious way, the quantum theory. However, quite soon the two great men were to come into conflict.

The first round went to Einstein. In 1924, Bohr, Kramers and Slater (BKS) published a paper on the interaction of radiation and matter. They were concerned that in the interaction between radiation and matter the energy in the radiation field varied continuously but in quantum transitions there were discontinuous changes. It seemed to them that to account for this it would be necessary to abandon strict conservation of energy, and to replace it with a statistical conservation law. They were determined to avoid the light-quantum. Actually Einstein had had the same idea as early as 1910, but abandoned it and never published it.

In fact, the conservation of energy in individual interactions was demonstrated by Compton in his studies of the interaction of x-rays with matter, in which energy conservation and the existence of photons were clearly demonstrated in individual events. Bohr recanted and said he would give the BKS ideas a decent burial.

Thus the stage was set for the onset of the new quantum theory, but there was one final act involving Einstein, which straddled the old and the new. In June 1924 Einstein received a letter from a young Indian physicist named Satyendra Nath Bose. He enclosed the manuscript of a paper which had been turned down by *Philosophical Magazine*. He asked Einstein to read it and, if he thought it suitable, to submit it to *Zeitschrift für Physik*. Einstein read and approved it, translated it from English into German himself, and sent it off to the editor with a recommendation. Bose's paper dealt with a new method of calculating Planck's quantum law of radiation. At that time Planck's equation was not understood, even though it had been formulated twenty-five years before, and was not to be understood until it was explained by Dirac in 1926.

Two new ideas entered into physics with Bose: particles with two states of polarization, and non-conservation of photons (which still had not been named). Neither of these was regarded by Bose as important. Many years afterward he confessed that he had not at all appreciated the significance of these new concepts. What was immediately clear to Einstein, however, was that Bose had taken the first steps in quantum statistics. He immediately followed with three papers of his own. Fermi–Dirac statistics were formulated in 1926. Bose–Einstein condensation was discovered in liquid helium by Keesom in 1928 but was not claimed as such until 1938.

7.3. Einstein and the new quantum theory

The statistical papers were the last published contributions of Einstein on the old quantum theory, but he continued to influence, and later to criticize, developments in the new theory. De Broglie, whose scientific career had been interrupted by the war, wrote his thesis in 1924, but had already published a paper in 1923 in which he suggested the existence of matter waves. De Broglie, questioned many years afterward, thought that Einstein probably did not know of the 1923 paper, but did have a copy of the thesis, which was sent to him for his opinion. Thus Einstein was able to discuss the possibility in September 1924, two months before the thesis was defended. Einstein, of course, gave credit to de Broglie for the idea in his correspondence.

Einstein also influenced Schrödinger, who referred to his 'brief but infinitely far-seeing remarks' [3]. Even Heisenberg recalled conversations with Einstein about philosophical positivism which helped to form his interpretation of quantum mechanics. Thus even as late as 1926 Einstein was by no means isolated from the mainstream of physics.

But in 1927, at the Fifth Solvay Conference, Einstein struck. He had been invited to give a paper on quantum statistics, but declined. He made only one brief comment during the discussion, but saved his intervention for breakfast next morning. The problem centred around the interpretation of quantum mechanics.

In the two years, 1926 and 1927, quantum mechanics (or wave mechanics) developed with extraordinary speed. Einstein's position was first of all enthusiastic, but became less so when Max Born and his school at Göttingen began to develop strange interpretations of the new physics. Einstein's reactions were always somewhat difficult to predict. Max Planck (quoted by Frank in Schilpp, 1949) remarked: 'roughly speaking, we may distinguish two conflicting conceptions in the philosophy of science: the metaphysical and the positivistic conceptions. Each of these regards Einstein as its chief advocate and most distinguished witness'. Uncharacteristically he remarked to Heisenberg: 'It is the theory which decides what we can observe'. Undoubtedly Einstein had been strongly influenced by the positivist Ernst Mach, but he forsook positivism early on for a realist view of nature. In this he was joined by Schrödinger and de Broglie, who both had misgivings about the direction in which the new physics was heading. (Schrödinger, private communication, 1952). Einstein approved strongly of Schrödinger's work, and took it that the waves were real. For Born, and later for Bohr, the waves were 'waves of probability'.

According to Pais, the chronology of quantum mechanics was as follows.

1923. De Broglie puts forward the wave particle duality for matter, in two papers in *Comptes Rendus*. They did not attract much attention. In 1924, however, Langevin, de Broglie's supervisor, sent a copy of de Broglie's thesis to Einstein for an opinion. The thesis was due to be defended in November, but Einstein was talking about duality, and giving the credit to de Broglie, two months before this. He was very much taken with the idea. It must be remembered that wave–particle duality of photons was accepted only with the Compton effect in 1923–4, so the whole idea of duality was still very new.

1925. Heisenberg's first paper appeared in July. It was recognized as highly abstract. Einstein objected to the probabilist tone immediately. Heisenberg was joined by Born. Einstein said that he took exception to every sentence in Born's paper and every sentence in a letter that followed it. Born, Heisenberg and Bohr proceeded with the probability interpretation and hailed Schrödinger's wave mechanics as evidence in its favour, when it was published early in 1926. Schrödinger was extremely annoyed about this unsought support, for like Einstein he had grave doubts about the probability interpretation. For him, the waves were real. It was shown that Heisenberg's and Schrödinger's formulations were mathematically equivalent. The controversy hinged on interpretation.

The uncertainty principle was published in March 1927. Einstein fastened on it, in an attempt to disprove it. From now until 1931, it was his principal endeavour in quantum mechanics. Rather little of this was published in journals; it took place by letter, mainly with Bohr, and occasionally at conferences. It was carried out with high seriousness: Paul Ehrenfest, in conversation with a colleague, said, in tears, that his head was with Bohr, but his heart was with Einstein. Every word uttered by Bohr was weighed and measured. Einstein, though deadly serious, delighted in the construction of machines that would disprove the uncertainty principle. Patiently, Bohr would knock these 'thought machines' down again.

The uncertainty principle states that the product of the uncertainties in appropriate pairs of variables, such as momentum and position, or energy and time, is equal to or greater than Planck's constant of action h. Thus one can have high accuracy in the measurement of momentum, but the accuracy in position measurement will be correspondingly poor.

Otto Stern recalls breakfasts at the Fifth Solvay Conference in October 1927 [4]:

> Einstein came down to breakfast and expressed his misgivings about the new quantum theory, every time he had invented some beautiful experiment from which one saw that the theory did not work ... Pauli and Heisenberg who were there, did not pay much attention 'ach was, das stimmt schon, das stimmt schon' [ah well, it will be all right, it will be all right]. Bohr on the other hand, reflected on it with care and in the evening, at dinner, we were all together and he cleared up the matter in detail.

The battle between Einstein and Bohr fell into two phases. In the first, lasting roughly from 1925 to 1930, Einstein sought to show that quantum mechanics, and in particular the uncertainty principle, was incorrect. In the later stage he accepted the logical correctness of the theory, but held that it was incomplete. He continued in this belief for the rest of his life. Rather little of this dialogue was published in the literature. Most authors rely on an article by Bohr in a volume edited in 1949 by P A Schilpp, which is a characteristically careful account of the Bohr–Einstein dialogue [5]. The general lines of the discussions were rather similar, involving diffraction patterns of matter waves. There was one, however, which momentarily defeated Bohr (figure 7.3). Einstein produced this at the Sixth Solvay Conference in 1930.

A box, suspended from a spring balance, contains radiation. It is closed by a shutter which can be opened and closed by a clock, carried in the box. Thus the shutter can be opened for a controllable time. By weighing the box with the spring balance before and after the shutter is opened, the mass of radiation which has escaped can be determined.

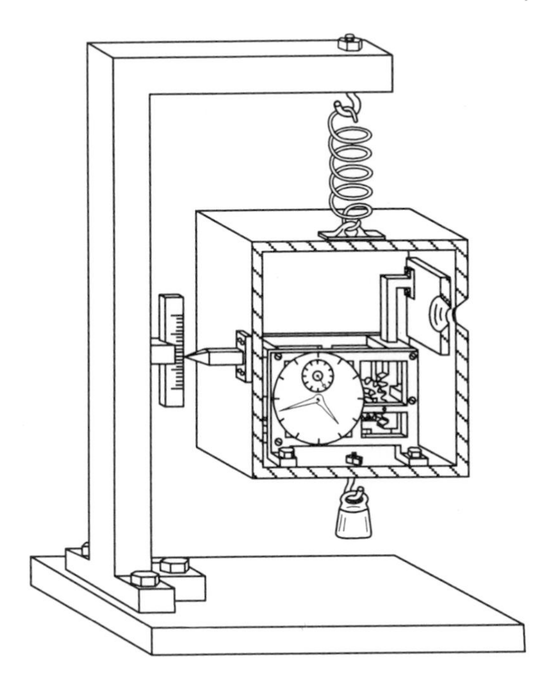

Figure 7.3. *The clock in a box.*

There is no limit to the accuracy with which the weighing process can be done. Hence the uncertainty principle does not hold.

Bohr was thunderstruck, and went around the gathering saying that it would be the end of physics if Einstein were right. Overnight, however, he worked out an answer, which most of those present accepted, and most physicists accept today. Einstein had forgotten the gravitational red-shift that a clock undergoes in a gravitational field. When this was taken into account, the uncertainty principle held.

The clock in a box was important because it marked a change in Einstein's attitude to quantum mechanics. Prior to this he concentrated on trying to demonstrate that the uncertainty principle was wrong. Afterwards, he accepted that quantum mechanics had no logical contradictions, but held that it was incomplete. This led to the Einstein, Podolsky and Rosen theorem, published in 1935 [6]. Here we will use it in a modified form ascribed to Bohm and Bell.

7.4. Einstein, Podolsky and Rosen

A pair of particles which have, reciprocally, spin up and spin down are emitted from a source into a vacuum. One (A) is directed on to a Stern–Gerlach magnet (a system which determines the direction of spin) close to the source. The other is directed to a similar magnet some considerable distance away. The direction of spin of A is determined. It is spin up.

We know immediately that B is spin down. This is apparently contrary to the principle of special relativity. Bell [7] points out, however, that a commonplace effect occurs often on a macroscopic scale. Suppose I have a pair of gloves, and inadvertently leave one at home. By looking at the one that I have and seeing that it is left-handed, I know immediately that the one at home is right-handed. No causal signal has passed. Indeed no signal has passed at all.

There is no problem here. For the microscopic system that we have described, however, things are not so simple. Bell [8] says:

> What is disturbing about quantum mechanics . . . is that before the first 'measurement' there *is* nothing but the quantum mechanical wavefunction—entirely neutral between the two possibilities. The decision between these possibilities is made for both of the mutually distant systems only by the first 'measurements' on one of them. There is no question, if there *was* nothing but the wavefunction, of just revealing a decision already taken. It was this spooky 'action at a distance', the immediate determining of events in a distant system by events in a near system, that scandalised Einstein, Podolsky and Rosen. They concluded that quantum mechanics must at best be incomplete. There must be in nature additional variables, not yet known to quantum mechanics, in both systems, which determine in advance the results of experiments, and which happen to have become correlated at the source—just as gloves are sold in matching pairs. It is now very difficult to maintain this hope, that local causality might be restored to quantum mechanics by the addition of complementary variables.

This last sentence requires explanation. Different authors use different expressions for what seems to be more or less the same thing. We have: local reality, local causality, locality, physical reality and so on.

Einstein, Podolsky and Rosen (EPR) started this plethora of terms:

> If, without in any way disturbing a system, we can predict with certainty (i.e. with probability equal to unity) the value of a physical quantity, then there exists an element of physical reality corresponding to that quantity.

In addition, they implicitly introduce locality [9]:

> If two systems have been for a period of time in dynamical isolation from each other, then a measurement on the first system can produce no real change on the second.

Most versions of the EPR thought experiment have involved spins measured in two directions at each point so four measurements are involved. When the paper first came out [10] it caused a flurry of excitement, but [11]:

Bohr as usual was imperturbable. The Copenhagen school had made a special point of emphasising that one ought never to think of quantum mechanical systems without also annexing to them the array of classical measuring instruments with which it was proposed to make the observations. It was, in their view, a package deal. Change the observations you were going to make and you had a new situation, even if the system to be observed remained the same.

In 1949 P A Schilpp invited a number of leading scientists to write critiques of Einstein's views, and Einstein to reply [12–15]. This is the main source of information about the two opinions.

So matters rested until after Einstein's death on 18 April 1955. In 1964, however, the Belfast-born physicist, John Bell, who was then working at CERN, propounded an inequality theorem which became famous. In it he states that a combination of numbers derived from the various spin orientations, should lie within a definite range, if a local theory holds. If the combination lies outside this range, quantum mechanics holds. Further predictions refer to the angular distribution of quantities (see Squires as listed in the Bibliography for details).

The experiments required to verify the theorem are very difficult, and a number of them gave ambiguous results. Recently, though, Aspect and his colleagues in France [16] have found very convincing evidence in favour of quantum mechanics. This raises difficult problems. It would appear that a measurement on a microscopic object A could affect the value of a quantity on a distant object B. Special relativity is not violated, but as John Bell said of the effect: 'It has a spooky feeling'. Fortunately Einstein did not live to see his worst fears come true.

7.5. Schrödinger's cat

At about the same time as the EPR paper appeared, Schrödinger put forward the following paradox: a cat is enclosed in a sound-proof box which also contains a gun, a radioactive source and a device for triggering the gun when a count occurs. If a count occurs, the cat will be shot. The source has an average rate of one count per hour.

The lid is placed on the box and left for an hour. We cannot observe what is happening inside the box, and according to quantum mechanics it is meaningless to ask such questions. After the hour we open the box. If the cat is dead then according to quantum mechanics we have killed it by opening the box.

Polkinghorne, however, makes the point that there is a conscious observer inside the box. It is the cat. It may not know if it is dead, but it certainly knows if it is alive. Its consciousness may not be very highly developed, but it is conscious. Thus the paradox may be spurious.

It would seem that the conscious observer is a necessary part of physics. The positivists have introduced the somewhat metaphysical concept of consciousness, but without defining it. Most physicists are probably, in practice, realists, though they may claim allegiance to the Copenhagen interpretation. Investigation seems somewhat pointless if there is no reality to investigate.

7.6. Einstein's personal life 1905–25

As we have said, Einstein's personality was extremely detached. His physicist friends were probably closer than his immediate family. Nevertheless he was surrounded from his early years with adoring family and friends. Except for his wife Mileva, however, who was not content to play nursemaid to genius. Their first child, Hans Albert, was born in 1904. Einstein was trying to get out of the Patent Office and into an academic career. But at that time it was necessary to serve an apprenticeship as a *privatdozent*, a person licensed to teach but not paid by a university. For most people an academic career was only possible if one had private means. He continued to work at the Patent Office while he taught a small class in his spare time. In 1909 he applied for an associate professorship at the University of Zurich. The Faculty report was favourable, but concluded with the following appalling passage [17]: 'These expressions of our colleague Kleiner, based on several years of personal contact, were all the more valuable for the committee as well as for the faculty as a whole since Herr Dr Einstein is an Israelite and since precisely to the Israelites among scholars are ascribed (in numerous cases not entirely without cause) all kinds of unpleasant peculiarities of character, such as intrusiveness, impudence, and a shopkeeper's mentality in the perception of their academic position. It should be said, however, that also among the Israelites there exist men who do not exhibit a trace of these disagreeable qualities and that it is not proper, therefore to disqualify a man only because he happens to be a Jew.' This kind of vicious anti-semitism was common in central Europe even before the First World War and even in liberal, neutral Switzerland. Einstein was appointed by ten votes to one.

In 1910 Einstein and Mileva had a second son, Eduard. As a young man he developed schizophrenia and had to be hospitalized. He died in 1965, still in hospital. Einstein did not stay long in Zurich. In 1911 he was appointed to a full professorship in Prague. He published little on relativity during his three years in Zurich, but continued to work on quantum theory, as we have seen. In 1912 he moved again, back to Zurich, but this time to the ETH. Still his wanderings were not over. Planck persuaded him in 1913 to go to Berlin, and he moved there to a research professorship in April 1914. Relations with Mileva were becoming impossibly strained, and they separated. Mileva took the two children back to Zurich. They were not formally divorced until 1919. Einstein's salary was not large, and the cost of maintaining his family in Zurich was a considerable strain on his resources.

The First World War was a difficult time for Einstein. A pacifist by temperament and conviction, he refused to be swept along in the wave of patriotic fervour. In 1915 and 1916 he completed his theory of general relativity; then in 1917 he was taken seriously ill. He had liver trouble and jaundice and a stomach ulcer. His cousin Elsa took care of him. In 1919, after divorcing Mileva, he married Elsa. Also in 1919, he swept into world fame with the results of the eclipse expeditions. His illnesses persisted until 1920. The rampant inflation which followed the First World War caused him further financial difficulties. He encountered anti-semitism again and became interested in Zionism. He was in great demand as a lecturer abroad. So, he lived as a world celebrity until 1925. In that year scientific certainties were swept away.

7.7. The later years

After the triumph of general relativity in 1919, Einstein travelled widely, lecturing to enormous and enthusiastic crowds, many of whom probably failed to understand a word he said about relativity. As we have said, he was interested in Zionism and was on the Board of the Hebrew University in Jerusalem, though he disagreed with its administration. He was offered, and refused, a professorship in Jerusalem, and later was approached about becoming the President of Israel. He left his manuscripts to the Hebrew University.

His pacifism changed with time. In the 1920s he was an ardent supporter of disarmament, but as the Nazi menace grew stronger, switched quite suddenly to support for rearmament against Germany. This disconcerted his disciples, but was characteristic of the man.

Another example of his sudden changes of attitude was his relationship with the League of Nations. He accepted an invitation to join the Committee on Intellectual Co-operation (CIC) in 1922. In 1923 he resigned from it. In 1924 he re-joined. In 1932 he finally resigned.

During the 1920s he went to many countries in Europe, North and South America and Asia. Since air travel was virtually non-existent he spent a lot of time on trains and ocean liners. He received honorary degrees in many countries, but only one from Germany, a doctorate in medicine from Rostock University. In England he received the Copley medal of the Royal Society, and the Gold Medal of the Royal Astronomical Society. Everywhere he went awards and honours were showered upon him, except in Germany. The stress took its toll and in 1928 he became ill again, being forced to stay in bed for four months.

In 1932 Einstein was offered a professorship at the Institute for Advanced Study in Princeton, New Jersey. Originally he planned to divide his time equally between Princeton and Berlin. He and his wife left for the USA on 10 December 1932. In 1933, after just one return visit to Europe, he sent in his resignation from the Prussian Academy and later from the Bavarian Academy of Sciences. On 17 October 1933

(Hitler was already in power) he, his wife, his secretary Helen Dukas and his assistant Walther Mayer arrived in Princeton. Apart from a brief visit to Bermuda to comply with American visa regulations for entry, he never left the United States again. Princeton, in fact, became the first real permanent home he had ever had.

His scientific work was mainly on the unified field theory, a Holy Grail of physics which many have sought. In modern terms this would correspond to the unification of the four forces of physics: electromagnetism, the strong and weak forces and gravitation. Einstein had no success, but in recent years physicists have achieved a partial unification of the forces. Einstein never lost sight, however, of the problems of quantum mechanics which we have outlined.

In 1936 his wife Elsa died, as well as his friend from student days Marcel Grossman. Einstein, always detached, became more so. His sister Maja arrived in Princeton in 1939, and made his house her home.

He made other contributions to the war effort, besides his famous letter to President Roosevelt warning him about nuclear weapons. In 1943 he became a consultant to the US Navy, not on nuclear weapons, but on 'high explosives and propellants'. This work was largely mathematical and similar work was done by other distinguished relativists such as J L Synge.

After the war he returned to pacifism, and with Bertrand Russell founded the Pugwash movement. He became, however, something of a grandfather figure, and was not active in peace demonstrations, unlike Russell. In 1945 he was sixty-six, and he slowed down considerably. An aortal aneurysm was discovered in 1948. He was forbidden to smoke, and agreed that he would not buy tobacco or matches. He had not promised not to steal them, however, and having an elastic conscience, became a familiar figure in the rooms of pipe-smoking colleagues, raiding their tobacco jars. Abraham Pais in his biography of Einstein records how Bohr, on one of his visits to Princeton, was wrestling with a problem in the interpretation of quantum mechanics. Pacing up and down as was his wont, he muttered 'Einstein, Einstein'. As if by magic, Einstein appeared, sneaking in to loot Pais's tobacco jar. They stared at each other for a moment, then burst into roars of laughter.

He died on 18 April 1955. He must be regarded as the greatest physicist of the twentieth century, certainly the most universally known. Yet he spent much of his life working in blind alleys. Would his theories have been discovered by others? Probably—that is the way with physics—but no one came near his breadth of achievement, and no one caught so strongly the imagination of people at large. He carried the aura of a nineteenth-century German professor, profoundly respected and with assistants at his beck and call, into the new world where graduate students regard their professors with ill-concealed contempt. It is improbable that we will see his like again.

Chapter 8

OPPENHEIMER AND THE AEC

In October 1953, Dr J Robert Oppenheimer was in London, giving the Reith Lectures for the BBC, on 'Science and the Common Understanding'. He was at the time Chairman of the General Advisory Committee of the United States Atomic Energy Commission. He received a message to the effect that his security clearance was being withdrawn. On returning to Washington this was confirmed by Admiral Strauss, Chairman of the Commission. So ended ten years of public service for the most famous physicist in the United States. Why did it happen? We must go back.

Julius Robert Oppenheimer was born in New York on 22 April 1904. His father was a textile importer who had built up a large business from small beginnings as an immigrant from Germany. His mother was American-born. The family was Jewish but not Orthodox. Robert went to the Ethical Culture School, a gentle enviroment for a frail child. Throughout his childhood he was often ill, did not play out and concentrated on intellectual pursuits. At the age of twelve he lectured to the New York Mineralogical Club, whose members were astonished and delighted by his precocity. His build was slight; even as an adult, though six feet tall, he weighed less than nine stone (126 lb).

He remembered himself as 'an unctuously good, repulsive child'. He was not without courage, however. At fourteen he was sent to summer camp. He was set upon savagely by the camp bullies, stripped and locked naked in an icehouse for the night. Nevertheless he insisted on staying in the camp until it finished, but never went again. He graduated from the Ethical Culture School top of his class, and then went to Harvard. He completed his courses, a mixture of classics and science, a year early. He was generally regarded as exceptionally promising.

In 1925 he went to Cambridge to join Rutherford's group, but Rutherford would not take him. Instead he worked with J J Thomson. Thomson was a Nobel Prizewinner and a former President of the Royal

Figure 8.1. *J Robert Oppenheimer (1904–1967).*

Society. He had resigned from the Cavendish chair to make room for Rutherford, the younger man, who had come down from Manchester just as Thomson had done forty years before. Thomson had, however, stayed on in the Cavendish, being allocated rooms by Rutherford. This did not work out very well. Rutherford, while good-hearted, was a natural imperialist of the kind that is found in most science departments. Thomson therefore found himself squeezed to more and more remote localities until he and his students were in a room called the Garage, hot in summer, perishing cold in winter. Thomson was sixty years old, and though he was the discoverer of the electron, remained by temperament a classical physicist. The few students he had had after Rutherford arrived found themselves on work which was not of the first importance. Oppenheimer discovered in Cambridge that he was a natural theoretician. Quantum mechanics was emerging but he could not contribute significantly, desperately as he tried. It is doubtful whether J J Thomson was able to help him very much. Oppenheimer did, however, manage to make some useful contributions in the field of applied quantum mechanics. The whole of spectroscopy, and the behaviour of high-energy electrons and gamma-rays, were wide open to investigation by the new methods. He published two papers on rotational and vibrational spectra, in the *Proceedings of the Cambridge Philosophical Society*. It was not the most fashionable journal, but it was a start. He realized that Cambridge was not the place for him. More congenial surroundings were found at Göttingen, where he was invited by Max Born. It was the centre of the new quantum theory, with

Heisenberg and Born, and the mathematician Jordan. The great Danish physicist Niels Bohr was a frequent visitor, as were Schrödinger and many others.

Oppenheimer desperately wanted to be one of the leaders in this field, but was not. Instead he wisely turned to the field of molecular physics, collaborating with Born. Their work was important and respected, but it was not in the vanguard. His doctorate was awarded in May 1927, and gained overall marks of excellent or very good. He was brilliant at developing other people's theories, but he was not a Dirac or a Heisenberg, and he was aware of this.

While in Cambridge he had problems, which his psychiatrist diagnosed as schizophrenia. After a holiday in Corsica his symptoms disappeared and never recurred, though he was to behave somewhat strangely at times. The Haakon Chevalier case during the Second World War will be described later. Göttingen had been good for him though he was never classed with Wigner, Pauli or Fermi.

Returning to America in 1928, he had a choice of several institutions. He chose the Berkeley campus of the University of California. This was a surprising choice since it was five years before Lawrence with his cyclotron made it into a world centre of physics. Oppenheimer, somewhat unconvincingly and rather preciously, said that his reason was that Berkeley had a very fine collection of sixteenth- and seventeenth-century French poetry. By and large it was a rather sleepy place at that time, unlike its present splendour. One may wonder whether he preferred to work in a place without the world figures that he would have encountered on the east coast. When Berkeley expanded in the 1930s, Oppenheimer commuted to Pasadena, in spite of the influx of Europeans all eager to work on the new cyclotrons. He himself maintained what must have been a somewhat lonely path in quantum electrodynamics, though he always had collaborators, and built up a good school.

His papers were mainly in *Physical Review* and numbered about 47 between 1928 and 1948. It is a respectable number considering what else he was doing, but it is not spectacular. Whatever the case, he realized that he was not going to make it among the leaders. However, he had other qualities. He was an inspired teacher, a brilliant negotiator, and, when the time came, a great leader. He evoked the loyalty of a wide variety of individualistic and difficult geniuses, at a time of great stress. Unfortunately, it seems that these qualities were not what he wanted. The fact that he had Enrico Fermi, the most brilliant and intuitive physicist of the century, working for him, was no consolation. He wanted to be Enrico Fermi.

His emotional life was equally confused [1]:

> In the autumn of 1936 I began to court Jean Tatlock and we grew close to each other. We were at least twice close enough to marriage to think of ourselves as engaged.

Jean was a communist, off and on, like many other intellectuals then, and drew him into fellow-travelling organizations at that time, the height of the Spanish Civil War, but he never joined the Party. She had a manic depressive temperament and committed suicide in 1944.

In 1939, Oppenheimer met the woman he was to marry. Her name was Katherine, but she was universally known as Kitty. She had been married to Joe Dallet, a Communist Party official who was killed fighting for the Republican cause in Spain. After that she married Stewart Harrison, a cancer research specialist at Caltech. She had also been a Party member, as was Oppenheimer's younger brother Frank.

Oppenheimer first got involved with the Bomb Project in October 1941, but momentous things had already happened. At Columbia University in New York, Enrico Fermi, a refugee from fascist Italy, and the Nobel Prizewinner in 1938, had constructed a number of assemblies to test the possibility of a nuclear reactor.

The project needed a large building; a disused squash court was found underneath a stand at Chicago University's stadium Stagg Field. There, at 3.48 pm on 2 December 1942, Fermi's team achieved a critical reaction. Leo Szilard said to Fermi: '... It is a black day in the history of mankind'.

Oppenheimer entered the project rather later than Szilard and Fermi. Compton was responsible for bomb development as well as the reactor work, and he appointed Gregory Breit to be in charge of fast neutron studies. Breit resigned in protest against what he considered sloppy security. Fermi in particular, he considered, talked too much. Fermi was unrepentant. Oppenheimer was appointed in June 1942, in place of Breit, and he set up a small theoretical group at Berkeley [2]. Thus when Oppenheimer arrived the project was in a situation of conflict.

He was appointed to be Director of a new bomb laboratory by General Leslie R Groves, who was overall director of the Manhattan Project, as he renamed it. Groves and Oppenheimer were a strange couple. The project cost a total of two billion dollars, so we are not talking small money. Groves was an engineer officer, who had been responsible for the building of the Pentagon at a cost of 10 billion dollars, so was accustomed to spending large sums of money. He was offered promotion to a one-star general if he took on Manhattan, the atom project, but he had been hoping for a posting overseas, with quicker promotion. His father had been an army chaplain, and he had been brought up on a succession of army bases. He was a West Point graduate with high Honours, and had also studied engineering at MIT, and at the University of Washington. He was a heavily built man, indeed overweight, with a somewhat unprepossessing appearance. He was also rather simplistic with a simplistic view of scientists, whom, on the whole, he disliked. For some reason though, he took to Oppenheimer, despite them being apparently poles apart: Oppenheimer, the Sanskrit-reading aesthetic philosopher, took up a friendship with the rugged general. This is important because of what happened later. Groves seems to have found some quality in Oppenheimer which inspired trust; unlike his experience with far more distinguished people, whom he just did not like. Groves complained only that Oppenheimer did not know much about sport.

Meanwhile, Oppenheimer was having a complicated life in Berkeley. He had a wife, a mistress, and a lot of friends, many of whom were communists. The situation was tortuous and we will not go into it in detail, since this work is primarily about people persecuted for their science, not for political or religious reasons. It was fashionable among intellectuals during the Spanish Civil War to join the Communist Party or a front organization for the Party. Oppenheimer was no exception, nor was his wife, who was a former Party member, although Oppenheimer was not.

From the beginning he was mistrusted by both Army Intelligence and by the FBI. They mounted a twenty-four hour surveillance on him, but there was no need. He himself went to the security people and made a statement. His case was handled by a Colonel Boris Pash. It was complicated. In 1943, a teacher of Romance languages named Haakon Chevalier, talking to Oppenheimer at a party, said that he had been approached by a man called Eltenton. Eltenton, a somewhat mysterious figure, was looking for information on American scientific work, for the Russians. Oppenheimer rejected this immediately, saying 'That would be treason'. For a long time Oppenheimer refused to give Pash Chevalier's name, but eventually did. It resulted in Chevalier's dismissal, and ruined his life. Oppenheimer had not only revealed Chevalier's name to the military (though reluctantly), but had invented a complicated scenario, with microfilm, planted messages and all the paraphernalia of a spy film. There was no need or reason for this; one is tempted to hark back to the psychiatrist in Cambridge who had diagnosed schizophrenia.

The Army continued to watch him but no overt charges were made. Groves first met him in October 1942, and he appointed him Director of the new weapons laboratory on 15 July 1943, at the age of thirty-nine, though effectively the appointment had already been made the previous autumn. It caused a sensation amongst the inner community. His main qualification was his American nationality. He had no Nobel Prize, but Prizewinners would be working under him. He was not even a member of the Academy of Sciences, and his scientific output was not large. His one strong point, which everyone recognized, was his capacity for leadership. He had built Pasadena up into one of the strongest theoretical schools in the country.

The new laboratory had to be in a remote site, in case of a disastrous accident. It must be west of the Mississipi, for political and demographic reasons, and far from an international frontier. Oppenheimer knew such a place. He and his brother had a small ranch in the Sangre de Cristo mountains of New Mexico. Quite near to it was a mountain bowl of a suitable size called Los Alamos. Before considering its fortunes we will briefly review the overall state of the Manhattan Project.

There are two ways of making a fission bomb: with uranium 235 or with plutonium 239. U235 was the first to be recognized, early in 1939. It occurs in nature as the rare isotope of uranium. It is mixed, however, with the U238 isotope, which constitutes over 99% of natural uranium. The two are almost impossible to separate chemically; physical means are necessary, and these are very difficult because the masses of the respective nuclei are so similar.

Four methods of isotope separation were seriously considered: centrifuging, diffusion, thermal diffusion and electromagnetic separation. Centrifuging was quickly abandoned. In diffusion, gaseous uranium hexafluoride is passed many times through screens, each containing millions of tiny apertures. The lighter isotope passes slightly faster through the apertures, increasing its concentration on the far side. Many successive transits are necessary, and considerable electric power must be used to drive the pumps. To make enough U235 for a bomb, the process must be on an enormous scale. Initial work was done at Columbia University, but the main plant was subsequently built at Oak Ridge in Tennessee, close to the Tennessee Valley Hydroelectric Scheme, which could supply the electrical power needed. According to Groves [3], diffusion was given only third priority in 1942.

Ahead of it was electromagnetic separation. This involved passing an ionized beam of uranium through intense magnetic fields. Because of the slightly different ratio of charge to mass, the two isotopes separated almost completely in one transit. The plant was also located at Oak Ridge, using enormous quantities of electricity to power the magnets.

An alternative method, which was given equal priority with electromagnetic separation, was the production of plutonium. This element does not occur in nature, because it has disappeared through radioactive decay. It can be produced, however, by bombarding natural uranium with neutrons. An intermediate element, neptunium, is formed and decays to plutonium 239. If the irradiation is continued for too long, another isotope is formed, plutonium 240, which tends to quench the chain reaction, and is therefore undesirable for weapons purposes. The neutron irradiation takes place in nuclear reactors, like Fermi's at Chicago, but much larger. These were situated at Hanford on the Columbia River in Washington State, a remote area with plenty of cooling water.

So far, however, there was no establishment devoted to the actual making of a bomb. Design and theoretical groups, including Oppenheimer's own at Chicago, were scattered throughout the country; and the British and French, who had been extremely active in design, had a laboratory in Montreal. Groves wanted a high-security weapons laboratory, run by the military, on a remote site. Hence, Los Alamos came into being. Oppenheimer went on a nationwide recruiting tour, and sent R R Wilson up to Harvard to negotiate the borrowing of their cyclotron. There was much reluctance. Few people wanted to bury themselves in a remote corner of the south west. Bethe, in particular, did not, though Oppenheimer was very keen to have him, and offered him the position of head of the theoretical division. Edward Teller, on the other hand, was keen to come.

Teller was a Hungarian, strongly anti-Communist. He was an abrasive but brilliant physicist, and was a strong proponent of the thermonuclear or 'hydrogen' bomb. This involves the fusion of two deuterium nuclei, or a deuterium nucleus with a tritium nucleus, at temperatures high enough to overcome the electrical repulsion of the nuclei. Only the explosion of a fission bomb, it seemed, could achieve this; but, unlike the case of the fission bomb, there was no limit to the size and hence the destructive power that could be achieved in a hydrogen bomb. The Los Alamos Review Committee, however, gave it low priority when it met in May 1943.

Work at Los Alamos had already started by that time. The site was operational but far from complete. Construction of temporary buildings continued throughout the Second World War, and development proceeded at a phenomenal rate. Accelerators were operating by mid-May 1943. An introductory course, given by Robert Serber, had started in April for the forty or so professional staff who had arrived. There was an air of extreme urgency, driven by the fear that the Germans would be first with the bomb. As it turned out, they were well behind. German nuclear progress, however, was not known to the Allies until nearly the end of the war in Europe.

Oppenheimer proved himself a magnificent leader right from the start at Los Alamos. He seemed to know everyone, even the construction crews, by name. He conveyed the immense urgency of the project without driving people. He took an intelligent interest in every experiment and theoretical programme, but he was not always right, and always needed convincing. Rhodes [4] refers caustically to: 'The stubbornness with which he rejected any possibility he had not himself foreseen'. This was not always liable to please some of the more temperamental of his staff, and he made some enemies. By and large, however, he was highly regarded.

Big names arrived as consultants: Fermi, Rabi and Allison came in April 1943. They were no longer needed elsewhere, because the other installations were turning into factories which could be run by more junior staff. Oppenheimer was assembling the most distinguished group of scientists who had ever worked in one place. This necessarily had a debilitating effect on Chicago, which was now involved, at long range, with the Hanford reactors and a smaller one in the Argonne Forest. Teller was assigned a group, not a division—he had perhaps expected more—and his housing was not good (most of the more senior scientists brought their families) so he had some cause for resentment. He had been in the project since the beginning, and Bethe, who got the Theoretical Division, was junior to him. This was Oppenheimer's decision, and one which was perhaps to prove significant. It may be relevant that Bethe had worked in the same area of physics as Oppenheimer, whereas Teller had not.

To crown the assembly, the great Niels Bohr arrived in December 1943. He was universally regarded with respect and affection, not only for his scientific ability, but also for his far-sighted wisdom. He had been smuggled from Copenhagen to Sweden, and flown from there to England in the bomb bay of an RAF Mosquito, nearly dying from oxygen starvation in the process. He had tried to persuade Churchill and Roosevelt to give information to the Russians, but was unsuccessful, annoying Churchill in particular. He did not stay to join the staff, but long enough to give them encouragement. He went to Washington in February 1944 [5], and renewed his struggle for international control of atomic energy.

As the months went by, and scientific and technical difficulties arose, friction also arose between some of the scientists. Their wives, too, complained about primitive housing, schools, mud and the all-pervading security checks. Some simply could not stand the barbed wire and fled. Also, serious scientific problems had arisen. The U235 bomb, it was thought, would not present a problem, provided enough sufficiently pure material could be obtained. One would simply take two sub-critical pieces of U235 and mount them at opposite ends of a cannon. When the cannon was fired the two pieces would join, make a critical mass and explode. There were of course many technical problems, but the scientists were confident that they could be surmounted. The main problem at the end of 1943 was lack of material on which to make experiments.

Plutonium, however, was more difficult. The simple gun method could not be used, because of spontaneous fission. Before the two halves of the bomb united, a stray neutron would initiate a partial explosion. One possibility was to simultaneously fire a number of sub-critical masses together, joining them in less than one tenth of a millionth of a second ($<1 \times 10^{-7}$ s). This was judged by explosives experts to be impossible. A strong Explosives and Hydrodynamics Division grew up, led, not like the others by nuclear physicists, but by experts on conventional explosives.

Teller, meanwhile, was pressing for more research on the 'Super', the thermonuclear weapon, to the increasing annoyance of Oppenheimer, who resisted this and in May 1944 advised Teller to withdraw altogether from fission work and form a thermonuclear group. From this point relations between the two men went steadily downhill, Teller becoming increasingly bitter.

By early 1945 most of the fission-bomb problems appeared to be solved, so much so that it was not considered necessary to make a test explosion of the U235 bomb. It could go straight into action. Plutonium continued to give problems. An implosion method was suggested by Seth Neddermeyer, in which a hollow sphere of plutonium would be compressed into a critical mass by an assembly of explosives outside the sphere.

Oppenheimer did not believe that this could be done, but agreed to research. An important advance was made by James Tuck, one of a small British contingent (another of whom was Klaus Fuchs, afterwards convicted of espionage). He had worked with shaped charges in designing conventional anti-tank weapons. These focus and intensify the explosion. Oppenheimer remained sceptical. The tests of the charges were largely under the control of George Kistiakowsky, a Russian-born Harvard chemist. There were personality and policy problems between him and Neddermeyer, which were temporarily resolved by Oppenheimer. The tests were successful.

But there was another, more far-reaching, emergency. The first Hanford reactor went critical and started to produce plutonium in September 1944. After a day's operation its activity declined and died. The problem was finally traced to an isotope of xenon, a by-product of one of the fission products. Modifications took until December, when secure operation was finally achieved. Meanwhile, there were also problems at Los Alamos. Neddermeyer and Kistiakowsky just could not get on with each other. A compromise was worked out. Effectively, Kistiakowsky would control implosion studies.

At this time, Teller and Oppenheimer were still on relatively good terms, in spite of Teller's bitterness over the Super. Rhodes quotes the former [6]:

> Oppenheimer was probably the best lab Director I have ever seen, because of the great mobility of his mind, because of his successful effort to know about everything important invented in the laboratory, and also because of his unusual insight into other people. . .

This was in stark contrast to the evidence which Teller was to give later.

In spite of his remarkable success at Los Alamos, Oppenheimer made enemies, some of whom surfaced later. Army Intelligence was still convinced that he was a spy. His wife found life at the site intolerable and began to drink heavily. He heard again from Jean Tatlock and went to see her in June 1943. This was duly recorded by Security Officers. In January 1944 she committed suicide. There was at least one spy at Los Alamos, Klaus Fuchs, working in the Theoretical Division under Hans

Bethe. He was quiet and unassuming, speaking only when spoken to. Possibly there were others.

The test of the plutonium bomb, 'Fat Man', was scheduled tentatively for mid-1945. Because of plutonium 240, and the spontaneous fission of plutonium 239 itself, implosion was the only possibility, and the Explosives Division were having a difficult time in developing an adequate system. Kistiakowsky recalls using a dental drill to remove bubbles in the plastic explosive primer. The shock waves from the different segments of explosive all had to arrive at the right place within a ten-millionth of a second. X-rays were taken of each segment to assess homogeneity and in test explosions ultra-fast photography was used.

In May 1945, the European War ended with the surrender of German forces. On VE Day, 8 May, a friend of the author's recalls driving down to Santa Fe to celebrate, with a multilingual group of Europeans. On their somewhat inebriated return to Los Alamos, they found most of their American colleagues grimly working away as usual.

As the war ended in Germany, the fear of what the Germans were doing receded. The German scientists had made some remarkable progress, but their hearts had not really been in it. Also the absence of experimentalists, mostly in forced exile and working in America, gave rise to difficulties. Harteck, the Hamburg chemist, protested to Heisenberg that a projected assembly, consisting of horizontal plates separated with heavy water, would be very inefficient. 'Ah yes,' replied Heisenberg, 'but with the assembly of small spheres you are suggesting, we could not use Bessel functions to solve the equations'. The complete divorce from reality implied by this reply is not, perhaps, typical, but the dominance of theoreticians in the German project, fortunately for the world, certainly was bad for the project. The remains of the German project, sabotaged a year earlier with the destruction in Norway by Norwegian commandos of most of the supply of heavy water, were mopped up by the ubiquitous Colonel Boris Pash of the ALSOS mission, sent into Germany with the advancing armies to look for signs of uranium research.

Meanwhile at Los Alamos preparations for the test of the plutonium bomb were proceeding at break-neck speed, but with many hitches. Oppenheimer was everywhere, sometimes helping, sometimes driving on, sometimes just interfering. The test was held successfully at Alamogordo, New Mexico, on 16 July 1945. It gave a yield equivalent to 18 kilotons of TNT. News was flashed to Truman who was in Potsdam meeting Stalin and Churchill (Churchill was replaced by Attlee halfway through the meeting as a result of the British General Election). Truman mentioned the bomb casually to Stalin, who wished him the best of luck. Of course Stalin knew quite a lot about it already from Soviet espionage.

Teller worked doggedly on with his theoretical group on the Super. They were somewhat left out in the preparations for Trinity, as the test was called. The uranium bomb was also ready, and it had already been decided where it should be dropped.

The President had set up what was known as the Interim Committee, to decide matters of policy. Parallel with this was the Target Committee. Oppenheimer was an adviser to the former and a member of the latter. So many of the Japanese cities had been wiped out that it was difficult to find a suitable target. The Target Committee studied seventeen cities, including Hiroshima and Nagasaki, for consideration by the Interim Committee. One of them, Kyoto, was vetoed by Henry Stimson, Secretary of War, on the grounds that it was sacred, and of great historical importance. Stimson was the key man; he would have to advise the President if and where the bomb should be dropped.

The deliberations of the Interim Committee have been described many times, notably by Stimson himself in an article in *Harper's Magazine* in 1947. The committee consisted of Stimson as chairman, Vannevar Bush, James Conant, Karl Compton, William Clayton, Ralph Bard of the Navy Department and a further representative nominated by the President. He turned out to be James Byrnes, whom Truman had appointed *in pectore* for the Secretary of State. There were no scientists on the committee, so a scientific panel was appended, consisting of three Nobel Prizewinners, Fermi, Lawrence and Arthur Compton, with Oppenheimer in addition. Such was Oppenheimer's standing at that time that he was nominated not only to the Target Committee, but also ranked with Nobel Laureates by the Interim Committee. He was only forty, two years younger than Fermi and much younger than the other two. This was probably his highest point in governmental influence, though of course he was quite unknown to the world in general.

The Interim Committee recommended use of the bomb on a mixed military and industrial target in Japan, without prior warning. It did so with a great deal of soul-searching and one dissentient (Ralph Bard), and the Scientific Panel, which did not have a vote, was sent away to consider alternatives. It met at Los Alamos, and reported back that it had been unable to find any.

At this time (June 1945), Leo Szilard in Chicago was circulating a petition against the use of the bomb in war. It attracted a fair number of signatories in the Chicago Laboratory, so he took it down to Los Alamos. Teller recalls that he wanted to sign it, but on consulting Oppenheimer was advised not to. Oppenheimer did not recall this conversation when asked about it after the war. Teller himself was totally occupied with his group working on the Super.

The Pacific War was reaching a crucial phase. As the Allies approached Japan the resistance of the Japanese became even more intense. The invasion of the Japanese mainland was scheduled for 1 November 1945, but the Joint Chiefs of Staff estimated in mid-June that the war would last for another eighteen months, until the end of

1946. A million more Allied casualties were expected, mostly in the landings. In these circumstances few people were keen to sign Szilard's petition and he returned to Chicago dispirited.

The bombs were duly used on Hiroshima on 6 August and on Nagasaki on 9 August. Six days later the war was over. Possibly it would have ended soon anyway, the point will continue to be argued, but in the West there was little sense of guilt, either among scientists or civilians, though a sense of revulsion crept in later as the scale and hideousness of the injuries became known.

After Hiroshima and Nagasaki, Oppenheimer became a national hero, even more so than Lawrence, Compton or Fermi. He was appointed Director of the Institute for Advanced Study at Princeton and seemed set to be a first-class scientific administrator for life. His physics output was not large, though one of his pre-war papers was to become very important with the rise of interest in black holes, but he was not a Nobel Prizewinner, and knew it. When the military relinquished control to the civilian Atomic Energy Commission (AEC), he became Chairman of the General Advisory Committee to the AEC, a very influential position.

Soon the question of the Super came up. Even as early as 1943 it had been discussed at Los Alamos. The basic nuclear physics was agreed. A mixture of deuterium and tritium must be heated up to some millions of degrees upon which the nuclei will fuse to form helium, releasing energy. There is no minimum critical size and no maximum size. As with the fission bomb, the reaction must be largely complete in a microsecond. An alternative is to use deuterium alone, which will also fuse to form helium but at a higher temperature than the deuterium–tritium reaction. Tritium, however, does not occur in nature, having a short radioactive lifetime. It must be made in a reactor.

In 1943 all the reactor capacity was in use producing plutonium. Tritium production would have been in competition with the race to produce a plutonium bomb and since a fusion bomb would in any case require a fission trigger to heat it up, the project was relegated to the background. Only Teller's group continued to work on it. After the war Teller retired in disgust to Chicago. Most of the scientists had already left Los Alamos to return to their universities. Oppenheimer, now on his way to Princeton, refused to get involved. Fermi was more helpful, but did not believe that the bomb could be made with deuterium alone, and also did not want to get involved. Like Teller he had returned to Chicago (Fermi, private communication, 1954).

Teller had told Conant, now back at Harvard, that it would probably take four or five years to build an H-bomb, but, strongly anti-Communist and anti-Russian, he felt it should be done. He believed, rightly as it turned out, that the Russians would have a fission bomb within five years and would then turn to the Super.

In April 1946 a conference was held at Los Alamos to discuss the

Super. Oppenheimer did not go and neither did many other senior people, but some of the younger ones went, as did Klaus Fuchs. Teller took the chair. They decided that a bomb was possible, but that enormous and complex mathematical calculations were necessary. Electronic computers hardly existed, so computations would have to be done on hand calculators, which in those days were mechanical devices. However, few people wanted to get involved.

Teller returned to Los Alamos in 1949 to engage again in weapons work. In September of that year the Russians exploded their first fission bomb. The General Advisory Committee (GAC) of the AEC met on 29 October 1949. Oppenheimer was in the chair and present were Conant, Fermi, Du Bridge, Rabi and others. The radiochemist Seaborg was not present. They recommended an increase in the pace of fission work, but were against the Super.

Rhodes says [7]:

> It based its recommendation on essentially two arguments: that at ten megatons a Super would be a weapon of mass destruction only, with no other apparent military use; and that it would not obviously improve the security of the United States. It would not do so, in particular, because the design then under consideration—the design reviewed at the Super conference in 1946, Teller's design, the classical Super—looked as if it would require large quantities of tritium, and plutonium and tritium, both created in nuclear reactors, would compete for existing plant capacity. Tritium was as well eighty times as expensive as plutonium ... The GAC members divided themselves into a majority and a minority to write explanatory annexes to their October 30 report. Conant drafted the majority annex, which Oppenheimer, Du Bridge, Rowe, Smith and Buckley signed. It said that a Super might become a weapon of genocide. It said such a bomb should never be produced. To the argument that the Russians may succeed in developing this weapon, we would reply that our undertaking it will not prove a deterrent to them. Should they use the weapon against us, reprisals by our large stock of atomic bombs would be comparably effective to the use of a Super. Rabi drafted the minority annex, which he and Fermi signed. It argued that the H-bomb question might serve as a springboard for new arms-control efforts. It described the Super as 'a weapon which in practical effect is almost one of genocide', then for good measure went even further in condemnation: 'It is an evil thing considered in any light'.

So the whole Committee, consisting of the most respected scientists and administrators in the country, were against the Super. Not just

Oppenheimer, though he was Chairman. The objections that they raised, however, were moral, not technical. This was a further blow to Teller.

On 31 January 1950, Truman announced that the Super should be made. It was a victory for Teller in particular, and a blow to the civilian atomic 'establishment'. But having decided that the bomb should be made; could it be made? In Teller's original design, using liquid deuterium and liquid tritium, it was enormously cumbersome. Oppenheimer had remarked that it could only be delivered by ox-cart. Nevertheless it was built and exploded, on 1 November 1952 at Eniwetok in the Pacific. In fact it was not a bomb that was exploded but a whole cryogenics laboratory. The small island on which it was housed, disappeared. Its power was equivalent to 10 megatons of TNT, 500 times that of the war-time fission bombs.

Between Truman's announcement in 1950, and the explosion at Eniwetok, had Oppenheimer (still Chairman of the GAC), continued to oppose the Super? He is not on public record as having done so but may have used his position to do so. We do not know. On his own account, he opposed it previously on technical grounds, not moral. (There is a contradiction here.) These technical grounds disappeared because of a number of inventions. It is not clear who made them, but Teller and Stanislaw Ulam were involved.

The first and most important one was over tritium. It, like deuterium, condenses to a liquid only at very low temperatures. Thus complicated and heavy equipment is required to keep the temperature down (the Eniwetok device weighed 65 tons). But how is tritium made? Normally in a reactor by bombardment of deuterium. Why not use the neutrons in an early stage of the explosion to make tritium? The problem of refrigerating the deuterium is solved by using a solid compound, lithium deuteride, with the mass 6 isotope of lithium.

Lithium-6 produces tritons (nuclei of tritium) by bombardment with neutrons. Lithium deuteride is stable, and solid at room temperatures. It was an ideal solution to the problem, and one which occurred also to the Russians.

However, that was not the only problem. Teller and Ulam came up with ideas that have never been fully de-classified, but which made a portable bomb possible and relatively simple. Oppenheimer described them as 'technically sweet'. A bomb based on these ideas was exploded in 1954. Its yield was larger than expected, at 15 megatons, and a number of Japanese fishermen in their boat the 'Lucky Dragon', were contaminated by fallout. It caused a world-wide protest.

The Russians had been through essentially the same routes, and were only a few months behind the Americans. The British developed a fission bomb in 1952 and a fusion bomb in 1956.

Oppenheimer's role in these developments, although he was still chairman of the GAC, was somewhat passive. Around the time of the first Russian and American hydrogen bombs, he was in London giving the Reith Lectures for the BBC, when he received notice that his Security Clearance for the Atomic Energy Commission was being terminated. He hurried home. It was the McCarthy time when half a nation was terrorized.

In fact his clearance cancellation had nothing to do with Senator McCarthy, but the predominant atmosphere in 1953 was one of fear. The Rosenbergs had just been executed for espionage, on very thin evidence.

Senator McCarthy's activities were primarily to do with suspected Communists. This book is concerned with scientific issues, but the fact that so much of the evidence presented in the Oppenheimer case was to do with his Communist connections, coupled with the factor that McCarthy was at that time at the height of his power, demands a digression.

McCarthy was born in 1909 in Wisconsin, traditionally a liberal state, and attended Marquette University. At the outbreak of war in 1941 he joined the Marines and afterwards became an air gunner. In 1939 he had been appointed a judge, so that when he ran for the Senate in 1945 he was acting unconstitutionally. In spite of a Supreme Court judgment he took his seat as a Republican in 1950, having gained a huge majority. His only policy was anti-Communism.

It should perhaps be said that anti-Communism in America was not simply a product of the Cold War, but went far further back, to the conflict between the oil, beef, railway and steel barons on the one hand, and the newly formed trade unions on the other, in the late nineteenth century. The union movement was populist, not Marxist. Governor Altgeld of Illinois, who had opposed the use of troops to break the Pullman strike of 1894, seemed the best hope for a constitutional social democratic party, but he was excluded from the Democratic nomination for President in 1896, because he had been born outside the United States. The nomination went to William Jennings Bryan, who lost to the Republicans in 1896, 1900 and 1908.

Bryan was a populist, obsessed with monetary reform, who had been running on the Free Silver ticket since 1890. His Christian fundamentalism and pacifism made him deeply suspicious of Marxism and Socialism. He was not the man who was needed, and Social Democratic parties of the European kind did not form in America.

The power of the police, and of state officials, to act against the unions, was enormous. In 1915 the Sheriff of Bisbee, Arizona, ordered the deportation of 1500 miners from the area, solely because they were members of a union. They were taken away in cattle trucks, with no possibility of appeal.

Thus a constitutional socialist party did not form and anyone with left-of-centre sentiments was more or less forced into the Communist Party. This gave rise to a deep fear of Communism on the part of the right wing of the Republican Party, but during the 1930s, and particularly during the Spanish Civil War, many intellectuals, like Oppenheimer, turned leftward.

The execution of Ethel and Julius Rosenberg for spying on the atom bomb project, coupled with the arrest and conviction of Klaus Fuchs in London, with Nunn May and Alger Hiss as minor figures, and with the war in Korea, gave McCarthy his chance.

In January 1953 he became Chairman of the Permanent Subcommittee on Investigations, and with the enthusiastic help of J Edgar Hoover and his FBI, launched a crusade. He attacked the media, the film industry in particular, and liberals of all kinds. No one was safe. He then turned on the government. He accused the State Department of having 205 card-carrying Communists, but was unable to substantiate this before the Senate Foreign Relations Committee. Finally he overreached himself by attacking the army and President Eisenhower himself. At this, even extreme Republicans withdrew their support, and with the revelation of financial irregularities, McCarthy fell from grace. He died in 1957.

On 21 December 1953 Oppenheimer was summoned to Washington by Admiral Strauss, chairman of the Atomic Energy Commission. There he was handed a letter drafted by General Nichols, General Manager of the Commission. It contained five charges (as reported by Associated Press):

1. Dr Oppenheimer had been associated with Communists on a number of occasions at the beginning of the war. He had had a Communist mistress and his wife was a former Communist. He had contributed generously to Communist funds from 1940 to 1942.
2. He had engaged Communists or ex-Communists to work at Los Alamos.
3. He had made contradictory depositions to the FBI about his attendances at Communist meetings at the beginning of the war.
4. He had rejected a suggestion that he should pass information to the USSR but had failed to report the approach to him for several months (the Chevalier–Eltenton affair).
5. He had strongly opposed the hydrogen-bomb project in 1949, when he was Chairman of the General Advisory Committee of the AEC. He had continued to campaign against the project even after President Truman's decision ordering the Commission to conduct research with a view to executing the project.

He could resign or be subjected to a review.

We are mainly concerned with charge 5, as this is the only scientific charge, but clearly it cannot be separated from the other four charges. If he did oppose the construction of a hydrogen bomb, could it be supposed that he may have done so partly because he had Communist sympathies?

Was he a Communist? It seems most unlikely. He was an intensely ambitious man, and it did not require his very high intelligence to see that Communism, however secretly held, would be an enormous handicap to advancement in the America of the 1940s and 1950s. The risk of blackmail alone would rule it out. It seems safe to assume that, after a juvenile flirtation, he had put it behind him by 1940. He could not of course dissociate himself from the subject completely, because of his family connections, but he was quite open about these.

In the short period between the test at Alamogordo and the use of the bombs at Hiroshima and Nagasaki, interest in a thermonuclear weapon grew; but as the war ended and scientists returned to their laboratories interest declined except in a dedicated few. Indeed governmental interest waned. In 1947 the number of nuclear weapons stockpiled was less than ten. Then in August 1949 the Russians exploded their fission bomb. It came as an enormous shock. Some observers had not expected it until 1965. Their bomb was followed by the Korean War. Under these pressures the campaign to make the thermonuclear bomb became stronger. Its leader was of course Edward Teller. Ernest Lawrence, who had won a Nobel Prize for his invention of the cyclotron, also favoured it. With Luis Alvarez, they appeared before the Congressional Joint Committee on Atomic Energy to urge that the bomb be made.

On 31 January 1950, as we have said, President Truman announced that work should go ahead to build the bomb. Three days later Klaus Fuchs was arrested in London, reinforcing the case. It was known that he had attended some of the seminars on thermonuclear weapons at Los Alamos during the war, and indeed as late as 1946.

Not everyone believed that the bomb could be built. Many feared that the bomb could be built but believed that it should not. Hans Bethe, a pioneer in thermonuclear reactions in stars said in an article published in 1950 [8]: 'If we fight a war and win it with H-bombs, what history will remember is not the ideals we were fighting for, but the methods we used to accomplish them. These methods will be compared to the warfare of Genghis Khan, who ruthlessly killed every last inhabitant of Persia'.

In reply to Nichols' ultimatum Oppenheimer sent a long, autobiographical letter. Reid says of this [9]:

> It was an unusual document written in great detail and at great length and with the considered eloquence with which Oppenheimer had so often impressed and powerfully influenced scientists and politicians alike. It was an autobiography showing what he considered were his achievements, his loves, his hates and his motives.
>
> In a style giving the periods it described an air almost of unreality, it told of the Jewish child born to a family which had him benefit

from the privileges of wealth. The young Oppenheimer was to pass through the best American and European universities on his way to what promised to be an untroubled life of academic physics. But the events of the late 'twenties and 'thirties, in which human lives were contorted by economics and politics, were to force him into living with reality.

The depression, the sufferings of Jews in Germany and the Spanish Civil War, all encouraged him to attune his political consciousness. He attached himself to the left-wing cause, serving on committees to aid the Spanish Loyalists. As with many other of his west coast academic friends, his interest in communism ripened and he quickly aligned himself with the party's humanitarian aims. But this love affair was brief. Two factors contributed to the beginnings of a disenchantment with communism. The first was the confirmation, by physicists he could trust who had lived through the Russian purges of 1936–9, of the rumours he had heard of the tyranny of the political trials, and the treatment meted out to the victims. The second was the Nazi-Soviet pact which apparently attempted to unite his new-found love of communism with his new-found hate of fascism. By the time he was offered Los Alamos and the role that was to change mankind's future as well as his own, he considered communist fellow-travelling to be a part of his past.

The portion of his letter which dealt with the war years was told with suitably muted pride: the pride of a man who had been stretched to his limits and who had emerged as a success, and as an international figure into the bargain ... 'I had become widely regarded as a principal author or inventor of the atomic bomb, more widely, I well knew, than the facts warranted. In a modest way I had become a kind of public personage. I was deluged as I have been ever since with requests to lecture, and to take part in numerous scientific activities and public affairs. Most of these I did not accept.'

What he did accept were numerous posts as adviser on United States nuclear policy. It was in one of these, as chairman of the General Advisory Committee, that he joined in the Committee's unanimous opposition to the development of the hydrogen bomb. Oppenheimer's claim was that when, in spite of this opposition, President Truman announced his decision to proceed with the thermonuclear programme his own opposition ended 'once and for all'. Never, he said, had he urged anyone not to work on the hydrogen bomb project. The letter ended: '... I have reviewed two decades of my life. I have recalled instances where I acted unwisely. What I have hoped, was not that I could wholly avoid error, but that I might learn from it. What I have learned has, I think, made me more fit to serve my country'.

The AEC was grinding towards the question of Oppenheimer's security clearance. Nearly all the evidence was to do with his Communist associations, with which we are not primarily concerned, though it is difficult entirely to avoid them. The Personnel Security Board (PSB) began its sittings on 12 April 1954. It had no criminal investigation function. The most that it could do was to remove security clearance. This would not affect his employment at Princeton in any way, but it would be a personal disaster. Remember that this would be at a time when he was Chairman of the AEC Advisory Committee. Effectively it was accusing the chairman of the bank of embezzlement, though criminal charges would be waived.

The members of the PSB were three: Gordon Gray, President of the University of North Carolina, acting as Chairman, Thomas Morgan, late President of the Sperry Corporation and Dr Ward Evans, Professor of Chemistry at Loyola University. Only one, therefore, was a scientist, and there was no physicist.

After much damaging testimony had been produced on Oppenheimer's Communist associations, they came to the question of the H-Bomb. Had Oppenheimer deliberately held it up, prejudicing National Security? Naturally, most of the testimony on this point came from scientists, and was favourable. But not all was favourable. The Eltenton–Chevalier affair, in which Oppenheimer had lied to the security services (doing great damage to Chevalier) was the starting point at which the scientific testimony began. It was an unfortunate beginning.

Of the scientific witnesses, Hans Bethe gave a glowing account of Oppenheimer's integrity and judgment. Robb, counsel for the AEC, asked him whether he regarded himself as a good judge of character. On Bethe's replying 'Yes', Robb pointed out that Klaus Fuchs had worked in Bethe's Theoretical Physics Division in Los Alamos.

A host of distinguished physicists, as well as General Groves, paid tribute to Oppenheimer's scientific abilities and to his integrity, but one witness gave a damning indictment. Edward Teller, who was sometimes called 'The Father of the Hydrogen Bomb', had already been mentioned in previous testimony as difficult to work with.

Robb asked him [10]:

> Do you or do you not believe that Dr Oppenheimer is a security risk?

Teller replied:

> In a great number of cases I have seen Dr Oppenheimer act—I understood that Dr Oppenheimer acted—in a way which for me was exceedingly hard to understand. I thoroughly disagreed with

> him in numerous issues and his actions frankly appeared to me confused and complicated. To this extent I feel that I would like to see the vital interests of this country in hands which I understand better, and therefore trust more.
>
> In this very limited sense I would like to express a feeling that I would feel personally more secure if public matters would rest in other hands.

This was the kiss of death. In spite of the glowing testimonials produced by distinguished scientists, soldiers and statesmen, the Teller testimony hung over the enquiry. In the McCarthy era it was enough. Oppenheimer's position had already been undermined by the evidence of his Communist associations and by the bizarre Eltenton affair. The Board's Counsel, Roger Robb, returned to it again and again, introducing into the record an interview which Oppenheimer had had with Colonel Lansdale, the Manhattan Project Security officer, in 1943 [11] (in the following, L = Lansdale, O = Oppenheimer, R = Robb):

L: Well now you see what you stated that he contacted, I believe it was three persons on the project, and they told him to go to Hell in substance.
O: Although probably more politely.
L: And how do you know that he hasn't contacted others?
O: I don't. I can't know that. It would seem obvious that he would have.
L: If you heard about them they unquestionably were not successful.
O: Yes.
L: If you didn't hear about them they might be successful or they might at least be thinking about it, don't you see? Now you can therefore see from our point of view the importance of knowing what their channel is.
O: Yes.
L: I was wondering is this man a friend of yours by any chance?
O: He's an acquaintance of mine, I've known over many years.
L: Well do you—I mean there are acquaintances and there are friends. In other words, do you hesitate for fear of implicating a friend?
O: I hesitate to mention any more names because of the fact that the other names I have do not seem to be people who were guilty of anything or people who I would like to get mixed up in it, and in my own views I know that this is a view that you are in a position to doubt. They are not people who are going to get tied up in it in any other way. That is, I have a feeling that this is an extremely erratic and unsystematic thing.
L: Here is, I want you to in no derogatory way understand my position again.

O: Well ... there is a very strong feeling. Putting my finger on it I did it because of a sense of duty. I feel justified...
L: Now, here is an instance in which there is an actual attempt of espionage against probably the most important thing we're doing. You tell us about it 3 months later.
O: More than that, I think.
L: More than that. When the trail is cold it's stopped, when you have no reason not to suppose that these cases that you hear about are unsuccessful, that another attempt was made which you didn't hear about because it was successful.
O: Possibly. I am very, very inclined to doubt that it would have gone through this channel.
L: Why?
O: Because I had the feeling this was a cocktail party channel. A couple of guys who saw each other more or less by accident
L: Well, people don't usually do things like that at cocktail parties...

Although somewhat incoherent, the recording is damning evidence. At the hearing Robb questioned Oppenheimer about it [12]:

R: Why did you go into such great circumstantial detail about this thing if you were telling such a cock and bull story?
O: I fear that this whole thing is a piece of idiocy. I am afraid that I can't explain why there was a consul, why there was microfilm, why there were three people on the project, why two of them were at Los Alamos. All of them seem wholly false to me.
R: You will agree, would you not, sir, that if the story you told was true, it made things look very bad for Mr. Chevalier?
O: For anyone involved in it, yes, sir.
R: Including you?
O: Right.
R: Isn't it a fair statement today, Dr Opppenheimer, that according to your testimony now you told not one lie to Colonel Pash, but a whole fabrication and tissue of lies?
O: Right.
R: In great circumstantial detail, is that correct?
O: Right.

Of all the evidence that was presented at the hearings there is really rather little evidence that he was a Communist. The most damning, the Eltenton–Chevalier affair, was freely volunteered to the security services by Oppenheimer himself, and seems to have been heavily embroidered in the telling. Was he a compulsive liar or fantasist? His reputation was somewhat unreal. Freeman Dyson (1993) quotes Philip Morrison, a very acute observer [13]:

> Morrison has known him well since he started working under him at Pasadena, their ages being approximately twenty and thirty. He

> said that at that time Oppenheimer was still exceedingly intense and aesthetic, and divided his leisure between reading St Thomas Aquinas in Latin, and writing poetry in the style of Eliot. He came of a wealthy American family, but went to Göttingen to study and became thoroughly Europeanised; for a long time he contemplated becoming a Roman Catholic but finally didn't; and to furnish his mind he learned to read fluently in French, German, Italian, Russian, Latin, Greek and Sanskrit. So that to some extent explains the sensitivity of his prose, and the awe in which he is held by even such close friends as Bethe.

Latin and Greek are reasonable, since his primary degree in Harvard was partly in classics. German is also reasonable since he spent two years in Göttingen. But one wonders, did anyone ever really test him out in Sanskrit, and how had he the time for all these pursuits in addition to his physics?

As an example of Oppenheimer's strange behaviour take the Eltenton affair. It began with a party at Oppenheimer's house. He was in the kitchen making Martinis with a devoted friend and admirer, Haakon Chevalier, who was a professor of Romance Languages at Berkeley. Chevalier mentioned that he had been approached by a man called Eltenton, who was looking for information on secret projects. Eltenton was an Englishman, a chemical engineer who had spent some years in Russia working for a British firm, and in America was employed by Shell Development. He was known to both Oppenheimer and Chevalier, sharing left-wing sympathies. Chevalier recalled that Oppenheimer seemed disturbed by the suggestion but did not really react strongly. Oppenheimer's recollection, 'that Eltenton told him that he had a method, he had means of getting information to Soviet scientists. He didn't describe the means. I thought I said "but that is treason", but I am not sure. I said anyway something "This is a terrible thing to do". Chevalier said or expressed complete agreement. That was the end of it. It was a very brief conversation.' But then he begins to introduce new people working at Los Alamos, codes, secret messages and a whole paraphernalia of espionage in which he becomes entrapped; and all apparently without foundation except for the brief conversation in the kitchen.

As already mentioned, Oppenheimer did not reveal the story for several months and then to a junior security officer, who passed it on to Colonel Pash. This delay was itself contrary to Official Secrets Regulations.

Apart from this strange episode, the Security Services could discover little more than guilt by association. He had associated with Communists and fellow-travellers: his wife, brother and mistress were all Communists or had been. However, Oppenheimer had never been a Communist Party member, and apart from contributions to left-wing causes during the Spanish Civil War period there was nothing to suggest he still had

Communist sympathies. He claimed that he had become disillusioned with Communism after the Moscow Trials of 1936, and severed his connections therewith. The FBI remained convinced of his Communism, but never produced any evidence to prove it. Visits to Jean Tatlock, carefully observed by agents, were freely admitted, but these were not criminal offences, even though she was a Communist.

It was true that he had employed ex-Communists at Los Alamos, but had he not done so, the staff numbers would have been distinctly smaller, and the records of these people were known to the authorities. His attendances at Communist meetings and his account of them to the security services were, characteristically, confused, but there were discrepancies in the evidence also.

Eventually he was vindicated by the tribunal to the extent that they found him not to be a security risk. He was convicted of bad judgment, not disloyalty, but also, though euphemistically, of telling lies. Whilst this is irrelevant to our main point of Oppenheimer's attitude to the hydrogen bomb, it clearly affected the attitude of the Board to him. They questioned him closely in an attempt to find out what had prompted this extraordinary behaviour; he could only manage the lamest of excuses and the most convoluted of explanations. The point which shocked them most was the injustice done to Chevalier, who was wholly innocent.

Colonel Lansdale himself was produced to give evidence to the Board. Colonel Lansdale made it clear that he did not like scientists, and Ward Evans, the only scientist on the Board, pursued this point.

Lansdale said [14]:

> The scientists *en masse* presented an extremely difficult problem. The reason for it, as near as I can judge, is that with certain outstanding exceptions they lacked what I called breadth. They were extremely competent in their field but their extreme competence in their chosen field led them falsely to believe that they were as competent in any other field.
>
> The result when you got them together was to make administration pretty difficult because each one thought that he could administer the administrative aspects of the Army post better than any Army officer, for example, and didn't hesitate to say so with respect to any detail of living or detail of security or anything else. I hope my scientist friends will forgive me, but the very nature of them made things pretty difficult.

Evans riposted:

> Do you as a rule dislike the scientific mind? Is it a peculiar thing?

To which Lansdale replied:

> I will say this, that during the war I came very strongly to dislike the characteristics which it exhibited.

Returning to Oppenheimer and the hydrogen bomb; he denied that he had ever opposed it after Truman's decision to go ahead with it in January 1950. He denied that he had attempted to persuade anyone not to work on it.

After much further discussion the Board retired to consider its report. It had been presented with three-quarters of a million words of evidence. In relation to the hydrogen bomb they said [15]:

> We cannot dismiss the matter of Dr Oppenheimer's relationship to the development of the hydrogen bomb simply with the finding that his conduct was not motivated by disloyalty, because it is our conclusion that, whatever the motivation, the security interests of the United States were affected.
>
> We believe that, had Dr Oppenheimer given his enthusiastic support for the program a concerted effort would have been initiated at an earlier date. Following the President's decision he did not show the enthusiastic support for the program which might have been expected of the chief atomic adviser to the Government under the circumstances. Indeed a failure to communicate an abandonment of his earlier position undoubtedly had an effect upon other scientists. It is our feeling that Dr Oppenheimer's influence in the atomic scientific circles with respect to the hydrogen bomb was far greater than he would have led this Board to believe in his testimony before the Board.

The final recommendation said [16]:

> In arriving at our recommendation we have sought to address ourselves to the whole question before us and not to consider the problem as a fragmented one either in terms of specific criteria or in terms of any period in Dr Oppenheimer's life, or to consider loyalty, character or associations separately.
>
> However, of course, the most serious finding which this Board could make as a result of these proceedings would be that of disloyalty on the part of Dr Oppenheimer to his country. For that reason, we have given particular attention to the question of his loyalty, and we have come to a clear conclusion, which should be reassuring to the people of this country, that he is a loyal citizen. If this were the only consideration, therefore, we would recommend that the reinstatement of his clearance would not be a danger to the common defense and security.
>
> We have, however, not been able to arrive at the conclusion that it would be clearly consistent with the security interests of the United States to reinstate Dr Oppenheimer's clearance, and, therefore, do not so recommend.

The following considerations have been controlling in leading us to our conclusion:

1. We find that Dr Oppenheimer's continuing conduct and associations have reflected a serious disregard for the requirements of the security system.
2. We have found a susceptibility to influence which could have serious implications for the security interest of the country.
3. We find his conduct in the hydrogen-bomb program sufficiently disturbing as to raise a doubt as to whether his future participation, if characterised by the same attitudes in a Government program relating to the national defense, would be clearly consistent with the best interests of security.
4. We have regretfully concluded that Dr Oppenheimer has been less than candid in several instances in his testimony before this Board.

The report was signed by Gordon Gray and Thomas Morgan. Ward Evans, the only scientist on the Board, did not sign it, but submitted a minority report. He said of Oppenheimer [17]:

> His judgment was bad in some cases, and most excellent in others but, in my estimation, it is better now than it was in 1947 and to damn him now and to ruin his career and his service, I cannot do it. His statements in cross examination have shown him to be still naive, but extremely honest and such statements work to his benefit in my estimation. All people are to some extent a security risk. I don't think we have to go out of our way to point out how this man might be a security risk.

Apart from the verdict itself, he was worried about its effects:

> His witnesses are a considerable segment of the scientific backbone of our Nation and they endorse him. I am worried about the effect an improper decision may have on the scientific development in our country.

The AEC took its vote on 28 June 1954. By four to one it agreed to accept the Gray report and deny security clearance to Oppenheimer. The dissentient was again a scientist, H D Smyth, author of the famous Smyth Report of 1945, which first broke the news of the atomic bomb to the world. The Commission stated that the opinions of Oppenheimer prior to January 1950 had not been taken into account, as debate before that was 'free' (Fermi and others had opposed the bomb prior to 1950).

Evans was correct in saying that the AEC decision split the country, at least as far as scientists were concerned. The country at large, and doubtless many scientists, were terrorized by Senator McCarthy, but there was still some debate. Oppenheimer supporters outnumbered opponents by a large margin, so free speech was not entirely dead, but the Government, and the AEC, had countless millions to spend on weapons, so the Teller group had the power. Development of the hydrogen bomb went on, without Oppenheimer.

There were no draconian penalties. Simply, he no longer had the right or the obligation, to see classified documents. His position as Director of the Institute for Advanced Studies was secure. Like Galileo, he was not torturable, nor was there any question of criminal charges. His friends, however, remarked a change in him. His activity in physics declined. In 1954, just before Fermi's death, Fermi summoned Teller (he still had that power) to ask him to try to heal the rift in American physics, but to no avail. The rift remained for years, but with a strong majority in favour of Oppenheimer.

The government finally made amends. President Kennedy, on the day he died, announced the award of the Enrico Fermi Medal of the AEC to Oppenheimer. But even that honour was somewhat spoilt, since Teller had received it the year before, and they did not restore his security clearance. It must be said though that Teller signed the nomination papers for Oppenheimer to receive the medal.

Oppenheimer died in 1967. He was not perhaps mourned to the extent that Fermi was mourned, but he had many friends who remained loyally by his side. He was a brilliant but intensely introspective man who never achieved the ambitions he had set for himself.

Freeman Dyson, who knew him well in his later years at Princeton, says [18]:

> His flaw was restlessness, an inborn inability to be idle. Intervals of idleness are probably essential to creative work on the highest level. Oppenheimer was hardly ever idle.

Oppenheimer's favourite poet was George Herbert, and his favourite poem 'The Pulley':

When God at first made man,
Having a glasse of blessings standing by,
'Let us,' said He, 'poure on him all we can;
Let the world's riches, which dispersed lie,
Contract into a span.'
So strength made first a way,
Then beauty flow'd, then wisdome, honour, pleasure;
When almost all was out, God made a stay,
Perceiving that, alone of all His treasure,
Rest in the bottome lay.
'For if I should,' said He,
'Bestowe this jewell also on my creature,
He would adore my gifts instead of Me,

And rest in Nature, not the God of Nature:
So both should losers be.
Yet let him keep the rest,
But keep them with repining restlessness;
Let him be rich and wearie, that at least,
If goodnesse leade him not, yet wearinesse
May toss him to My breast.'

Chapter 9

BLACKETT, CHERWELL, TIZARD. OPERATIONAL RESEARCH AND TOTAL WAR

9.1. Introduction

On 9 March 1945, a force of 334 B29 American bombers attacked Tokyo with incendiary bombs. It was the first use of the substance napalm. An area of 12 square miles was destroyed, killing 84 000 civilians and injuring 41 000 more. Many jumped into the Tokyo rivers to escape the heat. The rivers boiled.

It seems an almost obscene question to ask whether a war in which such a thing happened could be a 'Just' one. Yet this war, of all in recent times, had started with high and serious purpose, against a clearly evil regime, and under the obligations of solemn treaties. If ever there could be, in modern times, a Just War, this was it. For Guy Crouchback [1]:

> The enemy at last was plain in view, huge and hateful, all disguise cast off. It was the Modern Age in arms. Whatever the outcome there was a place for him in that battle.

Very many saw it so. The revelations afterwards: the Holocaust, the ill-treatment of prisoners of war and civilians in large numbers, all serve to reinforce the view that this war was Just. Yet our sense of its justice is flawed and uncertain. What went wrong?

Aerial bombardment of civilian populations is older than the aeroplane, since bombs were dropped from balloons, but in modern terms seems to date from about 1911, when the Italians bombed rebel tribesmen in Eritrea, and from the time of German Zeppelin raids in the First World War. The RAF bombed Kurdish villages in Iraq in the 1920s, but area bombing in the modern sense seems to date from the Japanese attack on Nanking in 1937. The American Government responded immediately [2]:

> This Government holds the view that any general bombing of an extensive area wherein there resides a large population engaged in peaceful pursuits is unwarranted and contrary to the principles of law and humanity.

This text is crucial in the light of Allied bombing policy in the Second World War. Let us review it also in the light of traditional teaching on the Just War.

Just War teaching in the Christian tradition derives mainly from Augustine, in the '*City of God*' and in '*Reply to Faustus the Manichean*' (quoted in Gill [3]). The treatment is diffuse and relies heavily on a literal use of the Old Testament. A more succinct discussion is given by St Thomas Aquinas. Aquinas says [4]:

> There are three conditions of a Just War. First, the authority of the sovereign by whose command the war is to be waged. For it is not the business of the private individual to declare war or to summon the nation. The second condition is that hostilities should begin because of some crime on the part of the enemy. Wherefore Augustine observes that a Just War is wont to be described as one that avenges wrongs, when a nation or state has to be punished for refusing to make amends for the injuries done by its people or to restore what has been seized unjustly. The third condition is a rightful intention, the advancement of good or the avoidance of evil.

These conditions would appear to be clearly satisfied. The authority of the British, French and later the American governments can hardly be disputed. The threatened attack on the Sudetenland, shamefully accepted by the British and French governments, and the overt attack on Poland were necessary grounds for a declaration of war and constitute the second condition. Later, the attacks on Pearl Harbour and on Singapore justified Allied involvement with the war on Japan. The rightful intention demanded by the third condition may also reasonably be assumed. While self-interest was obviously involved, the Allied decision to engage in war could not be described as imperialist or expansionist. We are left, therefore, with the condition that Aquinas does not mention in this passage, the justification of means.

The Second World War began with displays of gallantry on both sides which in some cases verged on the farcical. Cavalry charged tanks, British seamen in Montevideo attended the funeral of Captain Fritz Langsdorf, who had been their captor until the *Graf Spee* was scuttled on Hitler's orders. Later, the *Blitzkrieg* involved some attacks on civilian refugees, but by and large the Wehrmacht adhered to the normally accepted standards, and the Allies were impeccable if not competent. The bombing of Rotterdam, followed by Coventry and London, marked a decline in German standards, but the British, alone apart from the Commonwealth, could still claim the moral high ground. Crouchback and other gentlemen rested secure during the Battle of Britain and later during the war in the desert, perhaps the only environment in which a really Just War could be fought in modern times (due to the small civilian presence). The problem of non-combatants now arose with the decision to bomb German cities in 1941.

This problem seems to me a terribly difficult one. If the moralist is going to demand that not one innocent person should suffer, then no war in history would pass the test. What, then, is the value of Just War theory at all? Must the moral person stand by, watching tyranny in other countries, invasion of one's allies, possibly the brutal persecution of civilians and do nothing lest one innocent person be harmed by one's actions? The imagination recoils. Surely, on the principle of the lesser of two evils, it must be a duty to intervene on behalf of the weak and persecuted, even if some smaller number of innocent enemy citizens is harmed. This was the view of the American Council of Churches Commission in 1943. Speaking, not merely of bombing, but of saturation bombing, they said [5]:

> However repugnant to Christian feelings [they may be] such acts are still justifiable on Christian principles, if they are essential to the successful conduct of a war that is itself justified.

This statement does not give a licence to wholesale and indiscriminate slaughter. The war must itself be justified (leaving aside for the moment the question of civilians) but moreover the acts *must be essential, which means that they must be effective, a weaker requirement*. Was saturation bombing effective? P M S Blackett thought not.

Group Captain Leonard Cheshire VC, a humane as well as gallant man, defended Allied bombing policy in a television interview. It was, he said, the only way in which Britain could strike back at the Germans between 1941 and 1943. It was, of course, his own Pathfinder squadrons which helped to make British night bombing more discriminating during those years. Nevertheless, the degree of discrimination was not high.

If we accept effectiveness as an important moral criterion, we must examine Allied bombing techniques and their defects. The dictum that the bomber will always get through, was due originally to Trenchard, though it was taken up by Baldwin later. It influenced strategic thinking in the 1920s and 1930s to such an extent that Fighter Command was for a time starved of resources. The question of whether, having got through, the bomber would be able to hit its target, was another matter. Many strange and wonderful bombsights were developed, allowing for every conceivable factor. Presumably they were all tested, but in fact

Figure 9.1. *P M S Blackett (1897–1974).*

they did not work, under war-time conditions, even in daylight. Two solutions to the problem were found: the development of microwave radar, made possible by the cavity magnetron, and the guidance system Oboe. Neither of these were available in quantity before 1943, and a fierce argument developed as to their application. Even before this the strategic decision had been made. C P Snow treats the argument in '*Science and Government*' [6] and its appendix 'Postscript to science and government'.

9.2. The scientists

Snow mainly deals with Lindemann (Lord Cherwell) and Tizard, but more significant than Tizard in the later stages was Blackett.

Patrick Maynard Stuart Blackett was born in London in 1897. His father's family was Northumbrian; prosperous Yeoman farmers for generations. His mother was more aristocratic. He was destined for the Navy, and went first to the junior naval college Osborne, then to the senior college at Dartmouth. When war broke out in 1914 he was a midshipman and served at the Battle of the Falkland Islands. Later he saw action at the Battle of Jutland.

At the end of the war he resigned from the Navy as a Lieutenant and entered Magdalene College, Cambridge to study Physics. Rutherford was just in the process of moving to Cambridge from Manchester.

Figure 9.2. *F A Lindemann (1886–1957).*

Blackett joined him in 1921 and began to develop the cloud chamber, which had been invented by C T R Wilson. He was to be involved with cloud chambers for practically the rest of his working life, narrowly missing priority in the discoveries of the positron and the mu-meson. Indeed, when he was awarded the Nobel Prize in 1948, it was for his development of the cloud chamber. After working briefly in Göttingen he returned to Cambridge, then was appointed professor in Birkbeck College, London. In 1937 he became professor at Manchester. As a student, I was awestruck by his stately procession down the main stairs for lunch. He always walked in the dead centre of the staircase, disdaining the banisters. He held his hands, naval fashion, in his jacket pockets, with the thumbs protruding. He had no nickname; he was Professor Blackett. Besides his work on cloud chambers he was a founder of operational research and also did important work on geomagnetism. He was tall and strikingly handsome. The Tate Gallery has a bust of him by Jacob Epstein. Altogether he was a most impressive person.

F A Lindemann, later Lord Cherwell, was quite different. He was born in Baden-Baden of wealthy parents. His father was Alsatian but took British nationality. His mother was American. His high school, university and postgraduate education were in Germany, but he became fanatically British, particularly after the outbreak of war in 1914. Recruited by The Royal Flying Corps (RFC) as a civilian scientist (being born in 1886 he was rather old for active service, but more importantly, there was this slight doubt about his nationality). He showed a total disregard for his own safety. He was reputed to have been the first pilot to have deliberately put an aircraft into a spin and brought it out again. He became Director of The Royal Aircraft Establishment (RAE) at Farnborough. After the war, Tizard persuaded the electors at Oxford to appoint Lindemann Professor of Physics. Physics at Oxford then was in a terrible state. A previous professor had held the chair for fifty years, and it seems fair to say that during that period only one physicist of distinction, Henry Moseley, had been produced.

Lindemann's work had been quite prolific, by the standards of the time, before the war. Now he ceased publishing, but did build up the equipment in the laboratory, and after Hitler came into power in 1933, he rescued a number of Jewish physicists from Germany. Some of these settled in Oxford and built up an active low-temperature group.

His main activity, however, was courting the English upper classes at country houses up and down the land. He became very well connected, though his manner and appearance were not prepossessing. He was tall and heavily built. He lived on a diet of white of egg, Port Salut cheese and olive oil. He became a close confidant of Winston Churchill, who was attracted by Lindemann's physical courage.

The third important person in this saga was Sir Henry Tizard. Unlike the other two he was not physically impressive. C P Snow has described him as looking rather like an intelligent frog. He was a chemist, who had produced little, but nevertheless had become fellow of the Royal Society (FRS) and achieved a knighthood. He was a scientific administrator and a very good one. Good ones are rare and many scientists feel that we could do very well without them altogether. He too had served at Farnborough, in his case, as an RFC officer, and showed a courage equal to that of Lindemann.

He was born in 1885, so was slightly older than Lindemann. His father was a senior Naval officer, assistant hydrographer to the Admiralty, and also an FRS. Henry was intended for the Navy, but it was found that he had slightly defective vision so could not pass the medical examination. Henry was not particularly upset, but his father was. Instead of the Navy, Henry went to Westminster School and Oxford, where he studied chemistry. Finding no one to work under in Oxford when he graduated, he went off to Germany, to work under Nernst in Berlin. There he met Lindemann and they became friends. This then was the beginning of an interaction that lasted nearly forty years, beginning in moderate but not close friendship; ending in bitter enmity. In a review of Ronald Clark's biography of Tizard, A J P Taylor has described Tizard as a man who turned gold into dross. This is unfair; he should be remembered for a number of things, but particularly, with perhaps Air Chief Marshal Lord Dowding, as the man who made

Figure 9.3. *Sir Henry Tizard (1885–1959).*

the radar chain work in 1940. But we are diverging from the central struggle.

The argument went back to 1936. The Air Ministry set up a committee to consider air defence, with Tizard as chairman. Other members were Blackett and A V Hill. Lindemann was co-opted later. Their main preoccupation was, naturally, radar, or RDF, as it was known in the early days. Lindemann, a close confidant of Churchill, was brought in at the latter's behest, to the dismay of the other members. Lindemann was a physicist who liked gadgets, but who had achieved little himself. Tizard too only had a small body of published work, having spent much of his time in administration, but was generally regarded as having a far better mind. Blackett and Hill were of a different order, both being awarded Nobel Prizes in due course. Most of the committee saw radar in conjunction with a strong Fighter Command as the only possible defence against the bomber. The possibility that radar might one day become an offensive weapon could not then have been envisaged, but its defensive role was obvious, if it could be made to work. Lindemann obstructed the work of the committee atrociously, putting forward such ideas as aerial mines attached to long strings and dropped on the enemy bomber. He did not actually oppose radar development, but constantly put forward objections on points of detail. Eventually Blackett and Hill resigned in desperation and the committee broke up in disorder. To be fair, one of Lindemann's obsessions, infrared detection, was used successfully by German night fighters during the war, and had not been taken up by the British, but on the whole his presence on the committee was a disaster. Fortunately he was not at that time sufficiently influential to prevent the development and deployment of radar systems, which were to save Britain in 1940. Later on, however, he became very powerful indeed. When Churchill became Prime Minister, he took on Lindemann as his principal scientific adviser.

Early experience with bombing on both sides showed that estimates of target accuracy had been wildly optimistic, and attempts to hit military or industrial targets from high altitude were virtually hopeless. The Germans used an ingenious radio beam system to direct their aircraft in night raids, but the Allies soon worked out how to deflect it, so that the bombs fell astray. C P Snow takes up the story [7]:

> The row occurred in 1942, and it occurred over strategic bombing. We have got to remember that it was very hard for the Western countries to make any significant military effort in Europe that year. The great battles were taking place on the Russian land. So it was natural, and good military sense, that the Western leaders were receptive to any ideas for action. It is also true—and this was not such good military sense—that the English and Americans had, for years past, believed in strategic bombing as no other countries had. Countries which had thought deeply about war, like Germany and Russia, had no faith in strategic bombing and had not invested much productive capacity or many elite troops in it. The English had, years before the war began. The strategy had not been thought out. It was just an unrationalised article of faith that strategic bombing was likely to be our most decisive method of making war. I think it is fair to say that Lindemann had always believed in this faith with characteristic intensity.
>
> Early in 1942 he ... produced a cabinet paper on the strategic bombing of Germany ... The bombing must be directed essentially against German working-class houses. Middle-class houses have too much space around them, and so are bound to waste bombs; factories and 'military objectives' had long since been forgotten, except in official bulletins, since they were much too difficult to find and hit. The paper claimed that—given a total concentration of effort on the production and use of bombing aircraft—it would be possible, in all the larger towns of Germany (that is, those with more than 50,000 inhabitants), to destroy 50 per cent of all houses.

The figures were sent to Tizard, who replied that they were too optimistic by a factor of five. Blackett, a founder of operational research, independently found them optimistic by a factor of six. The moral issue was not involved, here. The argument was about strategy only.

If Blackett and Tizard were right, the proposed offensive would be ineffective, and hopelessly wasteful. The Chief of Air Staff requested clarification:

> ref the new bombing directive: I suppose it is clear that the aiming points are to be the built-up areas, not for instance the dockyards or aircraft factories where these are mentioned in Appendix A. This must be made quite clear if it is not already understood.

The details of the calculations are given by Snow, but are too lengthy to be enumerated here. I would like, however, to quote a passage at some length from an article by Blackett. It summarizes the whole argument and its outcome, very well [8].

> No part of the war effort has been so well documented as this campaign, which had as its official objective 'the destruction and dislocation of the German military, industrial and economic system and the undermining of the morale of the German people to the point where their capacity for armed resistance is fatally weakened'. Immediately after the war the US Strategic Bombing Survey was sent to Germany to find out what had been achieved. A very strong team (which included two men who are now advisers to President Kennedy, J K Galbraith and Paul Nitze) produced a brilliant report, which was published in September 1945.
>
> Without any doubt the area-bombing offensive was an expensive failure. About 500 000 German men, women and children were killed, but in the whole bombing offensive 160 000 US and British airmen, the best young men of both countries, were lost. German war production went on rising steadily until it reached its peak in August, 1944. At this time the Allies were already in Paris and the Russian armies were well into Poland. German civilian morale did not crack.
>
> Perhaps it is not surprising that the report of the Strategic Bombing Survey seems to have had a rather small circulation; it is to be found in few libraries and does not appear to have been directly available, even to some historians of the war.
>
> If the Allied air effort had been used more intelligently, if more aircraft had been supplied for the Battle of the Atlantic and to support the land fighting in Africa and later in France, if the bombing of Germany had been carried out with the attrition of the enemy defences in mind rather than the razing of cities to the ground, I believe the war could have been won half a year or even a year earlier. The only major campaign in modern history in which the traditional military doctrine of waging war against the enemy's armed forces was abandoned for a planned attack on its

> civilian life was a disastrous flop. I confess to a haunting sense of personal failure, and I am sure Tizard felt the same way. If we had only been more persuasive and had forced people to believe our simple arithmetic, if we had fought officialdom more cleverly and lobbied ministers more vigorously, might we not have changed this decision?

Blackett, as we have said, had been a Naval Officer and had served at the Battle of Jutland. This background perhaps pre-disposed him to emphasize the importance of the war in the Atlantic. Perhaps he also expressed the traditional naval distaste for involving civilians in warfare, a luxury not always available to the Army and Air Force, but there seems little doubt that he and his colleagues were right strategically, so that our criterion of effectiveness was not satisfied. We could rephrase our criterion as 'a reasonable chance of success'. We must take up later what we mean by success.

The Blackett and Snow conclusions have not passed unchallenged. David Irving [9] has analysed the results of seven major bombing raids, including the Hamburg and Dresden raids. He concludes that Cherwell's estimate of the number made homeless was within forty percent of the actual number; whereas Blackett's prediction of the number killed by a given tonnage of bombs underestimated the actual number by a factor of fifty. In these particular cases, however, fire storms were created; which had not been anticipated by either Blackett or Cherwell. Most of the death and destruction was caused by incendiaries, whereas both of the calculations referred primarily to high explosives. The agreement with Cherwell's estimate could therefore be fortuitous.

The losses suffered by the attacking forces were appalling. Frankland, co-author of the official history of the bombing offensive, says of March 1944 [10]:

> The German air defences had got on top of the night bombers and were inflicting an insupportable casualty rate on them. In March 1944 Bomber Command was no longer in a position to sustain a major night offensive against German cities. In fact the tactical conditions of daylight had invaded the night to the extent where the cover of darkness had been fatally compromised. The night bomber never had the capacity to fight. It now also had an inadequate capacity to evade ... In the thirty-five major actions between November 1943 and March 1944, German night fighters destroyed the majority of the 1047 British bombers which failed to return. The daily average of bombers available for operations in the front line during this period varied from just over 800 to just under 1000, so that within five months, the German air defences, and principally the German night fighter force, destroyed more than the equivalent of the whole front line of Bomber Command.

The Americans fared little better in daylight. On 14 October 1943, 60 out of 291 Flying Fortresses were destroyed in the famous raid upon the Schweinfurt ball-bearing factory. Only the arrival of the remarkable long-range fighter, the Mustang, which could accompany the bombers to their targets, saved the daylight offensive.

9.3. Who was responsible, and what did they intend?

In a Just War decisions must be taken by a lawful government. In Britain during the war the lawful government was the War Cabinet, under the Crown and in conjunction with Parliament. It would appear that we can take it that the effective government was the War Cabinet. As we have seen, Lord Cherwell submitted his proposal directly to the Cabinet as a paper. There was of course opposition, particularly from the Admiralty. They were desperate to get the heavy bombers, and later microwave radar, for the anti-submarine battle in the Atlantic. Nor, we can assume, were Fighter Command particularly enthusiastic, but they had no direct representation in the Cabinet.

It is generally assumed that British bombing policy was the creation of Sir Arthur Harris. This is not in fact the case. The notorious memorandum from the Chief of Air Staff, Sir Charles Portal, concerning the location of aiming points, was dated 14 February 1942. Air Chief Marshal Harris did not take up his duties as A.O.C. (Air Officer Commanding) Bomber Command until 22 February [11, 12]. Liddell Hart refers to a decision by Churchill as early as June 1941 [13]:

> The United States might pump in 'air' to keep Britain afloat but that would merely suffice to prolong the process, not to avert the end. Moreover the measure of this respite was offset by Churchill's midsummer decision to pursue the bombing of Germany with all Britain's puny strength.

There is, however, no direct mention of such a decision in either volumes 3 or 4 of Churchill's own history of the war. Indeed there is a cautionary message from the Prime Minister to the Chief of Air Staff, dated 7 October 1941 [14]:

> I deprecate however placing unbounded confidence in this means of attack, and still more expressing that confidence in terms of arithmetic ... Even if all the towns of Germany were rendered largely uninhabitable it does not follow that the military control would be weakened or even that war industry could not be carried on. The Air Staff would make a mistake to put their claim too high. Before the war we were greatly misled by the pictures they painted of the destruction that would be wrought by air raids...

It would appear that his opinion changed somewhat later on, perhaps as a result of Cherwell's paper of March 1942; though there is no mention of the paper in volume 4 of his history. Altogether he does not seem eager to claim credit or blame for area-bombing policy. Nevertheless it must be assumed that the Air Staff and Bomber Command proceeded with the approval of the War Cabinet.

In the joint document issued by Roosevelt and Churchill after the Casablanca Conference in January 1943 there is a clear reference to 'morale' bombing [15]:

> ... the directive to the Allied Air Forces ordered: 'the progressive destruction and dislocation of the German military, industrial, and economic system, and the undermining of the morale of the German people to a point where their capacity for armed resistance is fatally weakened.

It seems clear therefore that area bombing had the approval of government at the highest level. But what exactly did they intend? They intended the destruction of industrial capacity, that is clear, and this can be argued as a legitimate aim in war. But what did they intend in relation to the civilian population? They intended to destroy their morale by destroying their houses, not to destroy their capacity to make munitions by destroying their factories. Is this a legitimate war aim? Could any reasonable person believe that they would destroy civilian houses without killing a large number of civilians? It could perhaps be argued that it was the responsibility of the German Government to provide adequate shelters for civilians in industrial areas, or to evacuate women and children from industrial areas altogether. It should be said, however, that Blackett and Tizard, who opposed area bombing, did so on strategic, not moral grounds.

No shelters were adequate against the fire storms, and most of the German civilian casualties, whether in Hamburg, Kassel, Dresden or elsewhere, were victims of the fire storms in the later stages of the war. They died of asphyxiation or burns, not from high explosive. Was the fire storm predicted? It appears that it came as a surprise to both sides when it first occurred in Hamburg in July 1943 but, by the time of the Dresden raid early in 1945, it must have been well understood, even though the scale of the casualties was suppressed by the German authorities. If the fire storm was understood, could it be argued that it was a legitimate weapon?

There appear to have been no agreements in international law to guide policy. Sir Arthur Harris claimed that the only relevant restriction was an agreement dating back to the Franco-Prussian war of 1870, prohibiting the dropping of explosive devices from dirigibles. Bomber Command did not use dirigibles. The concept of the 'Open City' did not apply, since all major German cities were defended by anti-aircraft batteries; though in the case of Dresden, according to Irving, almost all of these had been removed for use against advancing Russian tanks [16].

We can attempt to apply the principle of double effect to the problem. A mediaeval concept, it has been restated in modern terms by Walzer [17]:

1. The act is good in itself or at least indifferent, which means, for our purposes, that it is a legitimate act of war.
2. The direct effect is in itself morally acceptable—the destruction of enemy supplies, for example, or the killing of enemy soldiers.
3. The intention of the actor is good, that is, he aims narrowly at the acceptable effect; the evil effect is not one of his ends, nor is it a means to his ends, and aware of the evil involved, he seeks to minimize it, accepting costs to himself.
4. The good effect is sufficiently good to compensate for allowing the evil effect; it must be justifiable under the proportionality rule.

Looking at these criteria, it is hard to see that any of them are satisfied by area bombing; and particularly by the use of the fire storm, if that was indeed intended. The act itself was the dropping of bombs on a variety of targets which included a high proportion of civilian houses. Incendiary bombs are perhaps more legitimate weapons, but most of the targets could not be described as legitimate. The direct effect, it appears, was mainly the destruction of civilian housing and the killing of civilians, with military or directly strategic effects being rather small until very near the end of the war. The intention was the destruction of the German war machine, which was legitimate, but the immediate intention included the destruction of housing, which is inseparable from the destruction of civilian life. Finally the good effect, as we have seen, was not proportionate to the evil effect. We conclude that the policy of area bombing was not justifiable under the principle of double effect. We must repeat, however, that it was always open to the German authorities to evacuate all civilians or at least women, children and those not working in munition factories from industrial cities.

We must be satisfied that the policy was ordered by the lawful authority, and not a wanton act of destruction by an unauthorized subordinate. Air Chief Marshal Harris has been widely blamed. He did not receive a peerage, unlike other officers of similar standing, and became something of a scapegoat after the war. Undoubtedly he carried out his orders with enthusiasm, but there seems no doubt that his orders came from above, and that the Cabinet was responsible. Allied bombing was popular with the British and American public. Only a few lonely voices were raised in protest, for example Richard Stokes, MP and Bishop George Bell of Chichester spoke out. The Archbishop of Canterbury and the Moderator of the Presbyterian Church expressed some anxiety but were reassured by the Government.

The moralist Elisabeth Anscombe, in her pamphlet on Mr Truman's honorary degree, says [18]:

> It was the insistence on unconditional surrender that was the root of all evil. The connection between such a demand and the need to use the most ferocious methods of warfare will be obvious. and in itself the proposal of an unlimited objective in war is stupid and barbarous.

If we accept, then, that Allied bombing policy was unjustified, because it was both ineffective and in its means immoral, does this compromise the whole justification of the Allied war effort? In retrospect it would appear not. This was a war which had to be fought. As Walzer says [19]:

> It was the paradigm, for me, of a justified struggle.

9.4. Non-combatants in war

While the sanctity of civilian life is emphasized in such modern textbooks as those by Gill [20] and Bainton [21], there are few explicit references to it in the earlier Christian or classical texts. Luther seems to take it for granted that war will be accompanied by murder, rapine and pillage, and yet justifies it. Aquinas demands a right intention on the part of the authorities, yet does not, in his most famous passage on war, specifically treat the rights of the innocent. Augustine sees the Israelite attack upon Canaan, which must have involved much suffering on the part of the inoffensive, as not only justified but morally imperative, since it was enjoined by Jahweh. Nor do the pre-Christian writers seem to pay much attention to the rights of women and children. This is important in the present context since war was made on women and children, and in the nuclear age will continue to be.

The point therefore is crucial. What are the rights of the innocent and is anyone truly innocent? Some moral theologians, driven snarling into a corner on this point, have settled on the unborn as manifestly innocent, hence the prohibition of nuclear weapons because of their genetic effects.

This is not immediately obvious. Why should a life mutilated in 2500 AD, be worse than one in 1989? Harmful mutations tend to die out, so our instinctive horror at the threat to human inheritance has an element of magical thinking about it.

The Second Vatican Council, it is true, issued a firm condemnation of saturation bombing [22]:

> ... with these considerations in mind the Council, endorsing the condemnations of total warfare issued by recent popes, declares: Every act of war directed to the indiscriminate destruction of whole cities or vast areas with their inhabitants is a crime against God and man, which merits firm and unequivocal condemnation.

The reformed tradition seems to be somewhat different, as we have already seen from the quotation of the American Council of Churches. Bonhoeffer says [23]:

> One must speak of arbitrary killing wherever innocent life is deliberately destroyed. But in this context any life is innocent which does not engage in a conscious attack upon the life of another and which cannot be convicted of any criminal deed which is worthy of death. This means that the killing of the enemy in war is not arbitrarily killing. For even if he is not personally guilty, he is nevertheless consciously participating in the attack of his people against the life of my people and he must share in bearing the consequences of the collective guilt. And, of course, there is nothing arbitrary about the killing of a criminal who has done injury to the life of another; *nor yet about the killing of civilians in war, so long as it is not directly intended but is only an unfortunate consequence of a measure which is necessary on military grounds.*

Again:

> The suspension of the law can only serve the true fulfilment of it. In war, for example, there is killing, lying and expropriation solely in order that the authority of life, truth and property may be restored.

These texts would seem to grant more freedom of action to the military. Again, however, they could not be used to justify strategic bombing, if we accept that it was ineffective. A measure which is ineffective cannot be essential.

The American campaign against Japan was somewhat different. Although there were, of course, large factories, an astonishing amount of industry, including that making the major parts of aircraft, was carried on in backyard garages in the middle of Tokyo and other cities. The industrial complex was therefore inextricably intertwined with the civilian. The American claim was that effective attacks on strategic military targets could only be made by area bombing of cities. A similar problem arises with the urban terrorist who hides in a block of flats. How to attack him? To this extent therefore the bombing of Japanese cities was strategically more effective than that in Germany. After the Battle of Midway, Japanese aircraft production fell off, and once the Americans were able to establish bases within striking range of the Japanese mainland, great damage was done to Japanese industry. By that time, however, it was clear to all but the Japanese that the war was over. How to convince them? To many, the atomic bomb seemed the only answer.

In some respects the twenty-kiloton bombs which were used on Hiroshima and Nagasaki were comparable to a major fire-bomb raid. Casualties were similar and damage to property was also about the same. Injuries to people were different, though not necessarily worse than those due to napalm. The question of genetic injury arises, but let us leave that for the moment. Were the atomic bombs effective?

9.5. The atomic bombs

The controversy about atomic bombs continues today. It can be argued that they were effective in ending the Second World War, and that without them the war would have lasted until the invasion of the Japanese mainland, scheduled for 1 November 1945. In the landings planned, about a million casualties were expected between Allied troops, Allied prisoners of war ordered to be executed if the landing took place, and Japanese troops and civilians. Even if the collapse took place before the invasion, a continuation of the American bombing campaign for another month after 15 August would have cost more Japanese civilian lives than were actually lost in Hiroshima and Nagasaki. Thus a case could be made for the atomic weapons that could not be made for saturation bombing with incendiaries or for the earlier bombing in Germany.

There is some evidence for this view. Certainly some attempt to open peace negotiations had been made by a section of the Japanese Government under the new Premier, Admiral Suzuki, but it was resisted by an important section of the cabinet and by the army. Suzuki himself is reported to have said after the war that surrender would not have been possible without both of the atomic attacks [24]. Even after the Emperor had made his recording of the surrender speech, a coup was attempted by young officers. It is of course true that the United States and Britain did not welcome intervention in the Pacific war by the USSR, and the Japanese surrender was particularly welcome for this reason, but the machine guns had already been set up in the prisoner of war camps. Freeman Dyson says [25]:

> The primary source is Robert Butow's book, *Japan's Decision to Surrender*. Butow is an American historian who spent several years in Japan soon after the surrender, examining Japanese state papers and interviewing extensively the surviving members of the wartime Japanese government. He came closer than any later historian to answering the crucial question, 'would Japan have surrendered if the atomic bombs had not been dropped?'. I asked Butow this question explicitly when he was visiting Princeton. Butow replied, 'The Japanese leaders themselves do not know the answer to that question, and if they cannot answer it, neither can I'. Butow went on to explain that the Japanese government in 1945 was delicately balanced between the civilian leaders, who were trying to open peace negotiations through the Soviet Union, and the military leaders who were preparing to defend every inch of Japanese soil with the same suicidal ferocity with which they had defended Okinawa. None of us can know, even with hindsight, which way the balance would have tilted if the bombs had not been used.

Probably the argument will never be resolved. It is fashionable now to condemn the use of the bombs, whereas the use of conventional bombing which preceded them is almost forgotten. A question that might be asked is whether the use of the bombs was more unjust than that of an equivalent tonnage of conventional high explosive.

This kind of question could not be asked of modern thermonuclear weapons. No conventional weapon could approach their destructive power, but in Hiroshima and Nagasaki the destruction *was* of the same order as a fire-bomb raid. The argument would seem to turn on the nature of the injuries sustained by the victims. Both napalm and U235 are horrifying weapons, but the effects of nuclear bombs are more long-lasting, extending even into subsequent generations. Do the genetic effects of radiation constitute a fundamental barrier to their use? The argument seems difficult to sustain. If all genetic effects were inherently immoral we would have to prohibit the use of x-rays in diagnosis, since hospital x-rays are the largest source of radiation, after the natural background. The latter we can scarcely reduce, except by moving the whole population into limestone caves deep underground.

9.6. The battle of the Atlantic

This battle was the longest-running, most important and least publicized of all those in the war in Europe. The casualty rate, mainly amongst civilian merchant seamen, was the highest for any group in the whole war, except for a small number of high-risk groups such as the Chindits in Burma. 40 000 Allied seamen, one in four of the total force, were killed.

The battle began almost immediately after the war began, with the sinking of the Athenia, and ended on 7 May 1945, the day before VE Day, when a squadron of Fleet Air Arm Grumman Avengers bombed a cruiser in a Norwegian fjord, the last action of the European war. At the outbreak of war Admiral Doenitz had only 26 operational submarines, but some of them were already on station in the Atlantic, so that in the first month of the war 26 British merchant ships were sunk by U-boats. They also sank the aircraft carrier *Courageous,* and a month later, the battleship *Royal Oak,* in Scapa Flow itself.

Altogether, between 1939 and 1945, 1200 U-boats were involved, 700 of which were sunk, 2700 Allied ships were sunk and 32 000 German and 145 000 Allied sailors were drowned. Almost all of the latter were civilians and volunteers. Though the capital ships were of psychological importance it was basically a submarine war. The *Tirpitz,* for instance, never fired its main guns in anger, but its presence in a Norwegian fjord had a profound effect on the planning of the Murmansk convoys.

Churchill, in *The Hinge of Fate* [26] gives two maps of the positions of the sinkings of merchant ships, at two phases of the battle. In the first phase, from December 1941 to July 1942, the sinkings are concentrated along the American seaboard. After the introduction of coastal patrols by the Americans, the centre of gravity shifts very markedly to mid-Atlantic, with a patch of dots off the coast of Venezuela. Comparing this with the map given by Arnold-Foster of the limits of long-range patrols by the Liberator bombers of Coastal Command, the correlation is remarkable. After the introduction of centimetric radar in 1943, the rate of sinking of Allied ships declined steadily, and the sinking of U-boats increased sharply. What the Allies took some time to realize was that, unlike the situation in the First World War, U-boats were firing from the surface, not from periscope depth. Hence, if an aircraft located a submarine and had time to reach it before it submerged there was a good chance of sinking it. This was enormously helped by the new radar systems. Thus in March 1943, 43 merchant ships were sunk in the first 20 days of the month, but not a single Allied ship was lost between 17 May and September 1943 in the North Atlantic. The Germans tried numerous gadgets, but none was effective. Also of course the surface escort vessels were increasingly successful. It was too late for the invasion that had been hoped for in 1943, but 1944 was possible.

This dramatic turnaround in the Atlantic War was due to a number of factors:

1. The number and efficiency of surface escort vessels increased considerably.
2. Nearly 400 Liberator bombers, with a range of 3000 miles, were acquired by RAF Coastal Command; together with a number of Lancasters.
3. Centimetric radar was acquired for the Liberators, starting in March 1943.

The Germans were not slow to realize that something had happened. In mid-1943 Admiral Doenitz wrote [27]:

> For some months past, the enemy has rendered the U-boat war ineffective. He has achieved this objective, not through superior tactics or strategy, but through superiority in the field of science; this finds its expression in the modern battle weapon—detection. By this means he has torn our sole offensive weapon in the war against the Anglo-Saxons from our hands. It is essential to victory that we make good our scientific disparity and therefore restore to the U-boat its fighting qualities.

Doenitz, then, saw the most important of these as radar. It had been a long hard struggle, with many nations taking part. The relevance of the battle of the Atlantic to an article on strategic bombing is that the campaign in Germany was eating up scarce radar resources, and the decision to press ahead with strategic bombing was very nearly fatal. How far it prolonged the war is a matter of conjecture and argument.

Chapter 10

THE BIG BANG VERSUS CONTINUOUS CREATION

10.1. The size of the universe

As we have seen in the section on Hypatia, speculation on the nature, origin and size of the universe is very ancient. For Aristarchus, the world was largely confined to the solar system, with a distance of three million miles to the Sun, though he was aware that the stars must be very distant, since they showed no detectable parallax (that is, their relative positions did not change as the Earth moved around its annual orbit). Essentially for him, as later for Copernicus, the universe meant the solar system. Giordano Bruno introduced the idea of an infinite universe, but he was ahead of his time and never had any evidence for his view. Galileo was interested mainly in the solar system, though he did believe that the Milky Way was a collection of an enormous number of stars. The first step to a better understanding was the measurement of the distance to a star.

Christiaan Huygens, born in 1629, is best remembered today for his work on the wave nature of light. He was also, however, the first person to measure the distance to a star other than the Sun. Making the assumption that Sirius had the same intrinsic brightness as the Sun, he compared the apparent brightness of the two objects. Using the inverse square law for brightness, he found that the distance to Sirius was two hundred thousand times the distance from the Earth to the Sun.

It remained to measure the distance from the Earth to the Sun. Copernicus and Kepler had believed in an Earth–Sun distance of about three million miles, though it was not necessary for their work to know this quantity. Given relative distances for the planets, Kepler was able to derive his three laws. The measurement of one distance to a planet, however, would put a scale on the whole solar system. This was done in 1672 by a French expedition, who measured the distance from the Earth to Mars by a parallax method using a terrestrial baseline between Paris

Figure 10.1. *Christiaan Huygens (1629–1693).*

and Cayenne in French Guiana. This gave a value of 90 million miles for the Earth–Sun distance.

Huygens' result was not very accurate. Sirius does not have the same brightness as the Sun, but it was within an order of magnitude. The scale of the stars closest to the Sun came out at a few light years. (Two hundred years later, Bessel finally made stellar parallax measurements in 1838 and obtained similar values.)

Our Galaxy was studied very carefully by William Herschel in the latter part of the eighteenth century. In particular, he looked at nebulosities and classified them according to their appearance. He believed that some of the nebulae were outside our Galaxy, but had no proof for this. There was no method of measuring their distances at this time, or indeed even the distance to the centre of the Galaxy. By convention, Herschel placed the Sun at or near to the centre of the Galaxy, as did later authors. It is not clear how far this reflected the anthropocentric view of the universe which had prevailed before Copernicus.

Herschel was particularly interested in faint extended nebulae, which he called planetary nebulae because they resembled the images of planets. He rejected the suggestion that they were agglomerations of stars. Such an enormous object as the great Orion nebula could not be composed entirely of stars. At about the same time as Herschel's work, a catalogue of nebulae was drawn up by Charles Messier. Messier, like many other eighteenth-century astronomers, was a comet-hunter. The fuzzy images of nebulae could sometimes be mistaken for comets, so the point of Messier's catalogue was to help to eliminate these objects from the search. The nebula M1 in the catalogue was later identified with a supernova or exploding star which erupted in 1054 AD, and is now known as the Crab Nebula, but such objects were rare. Many of the objects in the catalogue were in fact spiral nebulae—galaxies like our own.

William Herschel's son John, working in South Africa, studied the Magellanic clouds, but did not succeed in measuring their distances. He built up a catalogue of nebulae. The Earl of Rosse, working in the 1840s at Birr castle in the Irish midlands, observed nebulae from Messier's catalogue with a six-foot reflector, at that time the largest telescope in the world. For the first time, he recognized the spiral structure which some of them had.

An important development took place with the work of Sir William Huggins, who used spectroscopic techniques to distinguish between certain types of nebulae. He found that the planetary nebulae had only one or two emission lines, like a hot gas. Spirals, however, showed a continuum spectrum like the Sun and other stars, implying that they had many stars within them. The great Andromeda nebula also showed a star-like spectrum. A few years later, in 1866, Huggins discovered the optical Doppler shift. He found that stars moving towards or away from the Earth, at speeds of about twenty miles per second, had tiny shifts in their spectra. This foreshadowed the red-shift techniques current today for probing very large distances in the universe.

By 1900, telescopes had improved in speed and accuracy and spectroscopic techniques were developing. A vast amount of data had accumulated on individual stars. Indeed there was a risk of overkill; the data had got out of hand. A given star could be described in more than a dozen different ways, depending on the catalogue from which it was taken. Another problem had arisen. Geologists estimated the age of the Earth, from the study of fossils, at several billions of years, but astronomers estimated the age of the Sun at about 100 million years. The problem lay in finding where the fuel had come from to keep the Sun burning much longer than the astronomers had predicted. Ultimately the answer was found in nuclear reactions.

Practically all models put the Sun at the middle of the Galaxy, and no other structures were recognized as being comparable in size to the Milky Way. Also Man remained at the focus of the universe. This anthropocentric view was opposed by many, notably Eddington, a distinguished British astronomer. The problem regarding the Sun was solved by careful observation, which showed that it lies in fact well away from the centre of the Galaxy.

The idea that the spiral nebulae might be comparable in size and structure to the Milky Way, had in fact been put forward as early as 1852, by Stephen Alexander in New Jersey, but even the suggestion that the Milky Way was a spiral was not accepted until after 1900. Still there were no measurements of distance for star clusters or galactic nebulae, since parallax methods fell short.

The first step towards the accurate determination of stellar distances was taken by Henrietta Leavitt at the Harvard College Observatory in 1908. She studied variable stars in the small Magellanic cloud. Since they were all in the cloud their distances were all approximately the same. The stars showed a pattern in their variability, over periods of one day to one month. The brightest star had the longest period. Thus the period of the star gave a measure of its absolute brightness. By measuring the apparent brightness, and comparing with a star of known period and distance nearby, the distance of the further star could be determined. Similar stars were found in the main body of the Milky Way, some of which had distances that could be measured by an independent method. These types of stars were called Cepheid variables, after the constellation in which the first of them was found. Hertzsprung used Leavitt's results to determine the distance to the small Magellanic cloud and showed that it lay outside our Galaxy.

The next person to take on the Cepheid variables was Harlow Shapley, working at Mount Wilson in California. He specialized in measurements on the globular clusters, which are collections of stars scattered near but outside the galactic centre. Taking these as defining the galactic centre, he finally and indisputably moved the Sun to the outskirts of the Galaxy, an important psychological as well as astronomical shift. The problem of the spiral nebulae remained. Were they just one of the strange objects in the Milky Way, or were they very distant separate galaxies? Was the Milky Way a spiral nebula itself? Many astronomers believed that it was, but proof was needed. Stars were resolvable in some nebulae and Cepheid variables were detectable. Their study was taken up by Edwin Hubble at Mount Wilson in the late 1920s and 1930s. He found Cepheid variables in an irregular nebula, NGC6822. Using measurements on stars near the Sun and studied by Shapley, together with the relationship between brightness and period, Hubble found a distance of 700 000 light years for NGC6822. This put it definitely outside the Galaxy and showed that it was an object rather similar to the Magellanic clouds. The nebulae M33 and M31 were similarly measured.

Thus we had two techniques for measuring distances: stellar parallax could measure up to about 100 light years; Cepheid variables could pick up from these and extend the scale up to a few million light years—enough to identify the spiral nebulae as galaxies outside the Milky Way. There the measurements stopped, because the Cepheid variables in more distant galaxies were too faint to be detected. Another method was necessary. It was possible to use supernovae ('exploding stars') in distant galaxies as standard candles. However, the ultimate weapon in the astronomer's armoury was the red-shift. By 1929 Hubble had estimated the distances of about twenty galaxies. He found that the light from them was shifted towards the red end of the spectrum. The more distant the galaxy the greater the red-shift. He found a linear relationship between distance and red-shift. There was scatter in the points and Burbidge has pointed out [1] that Hubble relied on the four most distant galaxies which he had observed to establish the linear effect. Results from closer galaxies were not clear.

The red-shift could therefore serve as a distance measure. In fact it was more than this, it showed that the universe was expanding: the greater the distance the faster the recessional velocity. Eventually, if the linear relationship persisted to great distances, the recessional velocity would reach the speed of light. Beyond this there would be no possibility of observing anything, so this distance defines an observable universe. The effect would be the same in all directions, so it would appear that Man is back at the centre of the universe. This is not so; an analogy sometimes used is that of points on the surface of a balloon which is being blown up. All points on the surface are equivalent, all see the same expansion when they look over the surface. No direction or point is preferred.

For a linear relationship, $v = Hr$, where v is the velocity, r the distance, and H a constant known as the Hubble constant. In 1929, Hubble found $H = 500$ km s^{-1} Mpc^{-1} (1 Mpc = megaparsec, approximately three million light years). In 1950, Baade reduced this to $H = 260$ km s^{-1} Mpc^{-1}, and in 1958 Sandage found $H = 75$ km s^{-1} Mpc^{-1}. Burbidge [2] favours an even lower value: $H = 52$ km s^{-1} Mpc^{-1}. The differences arise because there were errors in distance measurements, and because the absolute magnitudes of the brightest stars in galaxies were revised. One can say that the observable universe is from 10 to 20 billion light years in diameter.

10.2. The age of the universe

The age of the universe is of the order $1/H$. There was an embarrassing period when the Earth seems to be older than the universe, but with re-measurement of H the discrepancy disappeared. The exact value of the age depends on the acceleration or deceleration of the expansion, which may be important at large distances. The age would appear to lie between 10 and 20 billion years (one billion is one thousand million). In steady-state cosmology the concept of age is different; we will return to this. The deceleration parameter q measures the extent to which the universe is speeding up or slowing down. In figure 10.2 observational measurements of the expansion velocity are plotted against visual magnitude for a selection of galaxies. At small distances the

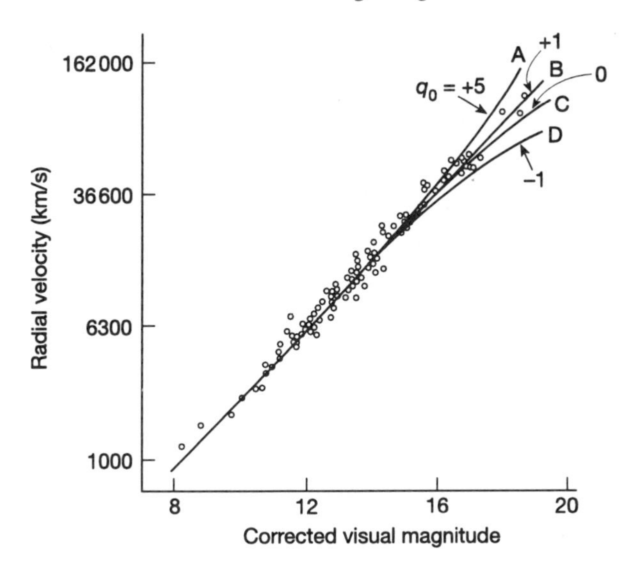

Figure 10.2. *Brightness versus red-shift.*

relation is a straight line but at large distances the line peels away above or below, depending on the value of q. q is also a measure of the characteristics of space.

If $q > 1/2$ the universe is spherical and closed; if $q < 1/2$ the universe is open. There is a special case if $q = -1$. This applies in the steady-state universe.

Everyone agrees that the universe is expanding, but not all agree that this is the result of one big bang. An expanding universe was inherent in Einstein's field equations of general relativity, but he found this undesirable, so inserted the cosmological constant to keep it static. He afterwards described this as the greatest mistake of his life. A particular type of expanding universe was predicted by the Russian relativist Friedmann in the early 1920s; this idea has since become very important. In 1928 the Belgian cleric Georges Lemaitre published his theory of the primeval atom, in which the universe originates in a very small, intense sphere which blows apart and expands uniformly. To Canon Lemaitre's profound embarrassment, Pope Pius XII took this up and put it forward, not quite as official teaching, but as something to be praised, in 1951. By the late 1940s, there was a general agreement about this kind of origin for the universe. This was challenged by Bondi, Gold and Hoyle.

In 1948 the accepted value of $1/H$ was only two billion years, less than the age of the Sun and Earth. This was soon remedied, but in the meantime it had contributed to a significant development. Bondi and Gold, as well as, independently, Hoyle, put forward what they called the perfect cosmological principle. Essentially this said that the universe was constant in time. The big-bang models predict universes which change their structure and size with time. The steady-state model says that the scale and structure of the universe remain the same indefinitely. This raises the problem of expansion. Since it is accepted that the velocity of recession increases with distance there must be a point at which it equals the velocity of light, or comes asymptotically close to it. Galaxies will be slipping over this boundary, and out of sight. To maintain the universe in a steady state the matter must be replaced. Hoyle introduced what he called the *c*-field. This causes the appearance of matter an atom at a time throughout the universe. Sciama [3] says there is a tension in the universe:

> In the steady-state model the density remains constant because the work done by the tension during the expansion results in a 'continuous creation' of matter. This just compensates for the diluting effect of the expansion.

Thus although this is often called the continuous creation model there is no breakdown of conservation, simply a conversion of energy into matter, an accepted process in physics. The appearance rate is 10^{-44} kg m^{-3} s^{-1}, or 1 atom per litre in a billion years, of cold neutral hydrogen. We will return to the model in detail later.

The expansion of the universe was established in about 1930, but there had been an outstanding anomaly for some hundreds of years. It is called Olbers paradox, but in fact goes back to Kepler and Halley. They had commented on the fact that the sky is dark, apart from stars, at night. In an infinite universe this would not be expected. Heinrich Olbers quantified this in the nineteenth century. Take a hollow shell of space surrounding the Earth. Stars inside it will contribute a certain amount of light to the night sky brightness. Now consider a shell of similar thickness but of twice the radius. The number of stars in it will be four times greater than in the smaller shell, but the contribution from each star will be only one quarter of the amount, because of the inverse square law. Hence the larger shell will contribute the same amount as the smaller one, to the night sky. Continue this throughout an infinite universe and the brightness will increase without limit. There are some defects in this reasoning. Light from a distant star may be cut off by a nearby star in front of it. This will indeed have an effect, and it may be shown that the resultant brightness will be equal to that of the Sun per unit area, so the whole sky will be as bright as the Sun. Another possibility suggested by Olbers is that there may be an absorbing cloud throughout the universe which reduces the light coming from great distances. This cloud, however, would be heated up until it was as bright as the illuminating stars.

Silk discusses the resolution of the paradox [4]:

Olbers' paradox can be resolved by modern theories of how the radiation originates. The most fundamental limitation on the total radiation is the finite age of the Universe. Stars do not live forever; their finite lifetime for producing radiation limits the resulting diffuse radiation density that stars can produce. Of course, it is unlikely that most of the matter in the Universe has already been processed in stars; the Sun, for example, which is still mostly hydrogen, seems to be rather typical of most stars in the Milky Way. Thus, stars cannot contribute more than a fraction of the Milky Way's brightness to the night sky.

We can also appeal to the redshift of the light from distant galaxies as an alternative means of resolving Olbers' paradox. The redshift amounts to a loss of energy, and the light from distant galaxies is highly redshifted. Its effective contribution to the night sky brightness is correspondingly diminished. However, this only accounts for visible light. Modern observations at longer wavelengths tell us that the sky is also dark even in the radio part of the spectrum. Therefore the redshift effect is not an explanation of Olbers' paradox. Nevertheless, the expansion of the universe plays an important role by limiting the volume of the universe within which stars are visible to us. It is precisely this fact, which restricts the observable universe to a distance of about 10 billion light-years, that results in a dark night sky: stars are inadequate radiators over this time scale. In this way, we can again reconcile the observed darkness of the night sky with theoretical expectations.

10.3. The radio universe

Until about 1940, the only source of information about the universe was gained from a very narrow band in the electromagnetic spectrum: visible light. It was only with the development of radar during and immediately after the Second World War that another channel opened up, radio astronomy. This is slightly surprising, since radio measurements on the Sun and on our Galaxy can now be done in a well-equipped undergraduate laboratory and radar itself goes back to 1904, so one might have expected that radio astronomy would have opened up earlier than it did. Some measurements were made by Jansky in 1932, and the work was followed up by Grote Reber in 1937, but it was the enormous expansion of equipment and personnel during the war which caused the growth of the new astronomy. Groups opened up in Cambridge and Manchester in Britain and in Sydney in Australia. Strangely, since the subject had originated in the United States, American groups developed more slowly. When they did develop, however, they built very powerful telescopes.

Apart from the Sun and our Galaxy, the first radio source to be identified was in the constellation Cygnus and was called Cygnus A. It was seen by Grote Reber in 1944, and confirmed by Hey, Parsons and Phillips in 1946. Further studies were made in 1948 by Ryle and Smith at Cambridge and by Bolton and Stanley in Australia.

There was some controversy about the mechanism by which these radio signals were generated. Ginzburg held that the radio flux from the Galaxy, at least, was largely due to synchrotron radiation—that is, radiation emitted by fast electrons spiralling in the magnetic fields of interstellar space. This has proved to be true though there are discrete sources also.

Signals from identified extragalactic sources such as the Andromeda Nebula, were discovered about 1950. Cygnus A, while very powerful, proved more difficult. Ryle and Smith showed that it was more than 2×10^{16} cm away, but this is only one-fiftieth of a light year, not as far as the nearest star. Baade and Minkowski found a faint member of a cluster of galaxies in the right place, and this optical identification was soon verified. In the early days catalogues proliferated, as they had done in optical astronomy around 1900, but it was soon appreciated that there were different kinds of source. The Crab Nebula and Cassiopeia A were strong sources, remnants of supernova explosions in our Galaxy. Cygnus A, however, was itself a galaxy, though a very peculiar one. Important catalogues were the 2C and 3C ones (short for Second Cambridge and Third Cambridge); these were followed by 4C and 5C. The last of these were used by Ryle and Pooley to conduct a radio source count, in 1968.

10.3.1. Radio source counts

It can be shown that if there is a uniform distribution of stationary sources extending indefinitely, a plot of the number N of sources whose radio brightness exceeds a value S will give:

$$N = S^{-3/2}.$$

If this is plotted on a log–log scale, as in figure 10.3 the resultant graph should have a slope -1.5. This situation would apply for the steady-state model if we ignore for the moment, effects of red-shifts. A correction for this would flatten the distribution. In fact, for the stronger sources the distribution is steeper than -1.5, and is quite a good fit to -1.8. For faint sources it will be seen that the distribution is flatter than -1.5 and has a progressive flattening as we go to faint sources. The law as we have described it ignores the variation of brightness from source to source. It also ignores the effect of red-shift. The variation of brightness does not in fact affect the basic -1.5 law. As for red-shift, Sciama has the following to say [5]:

> (i) The effective value of the brightness will depend on the redshift since we are observing in one small frequency-band radiation

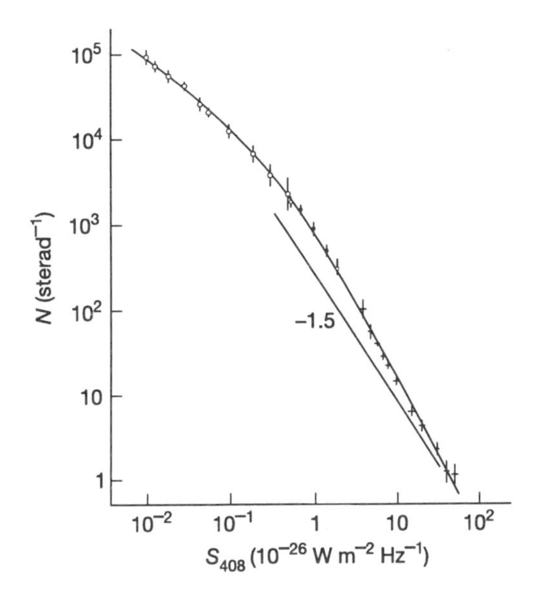

Figure 10.3. *Radio source counts.*

emitted in a different small frequency band. Allowance must therefore be made for the spectrum of each source.

(ii) The redshift reduces the apparent brightness of a source over and above the effect of the inverse square law, so that the sphere corresponding to S is reduced in size. This has the effect of reducing N.

(iii) If the redshift is taken to imply an evolutionary universe with no creation of matter, then the sources were more congested in the past than they are now, that is, the density was greater in the past. Now the greater the redshift the further the source, and so the longer we are looking into the past. Thus the effective density should increase with increasing redshift and so with decreasing S. This has the effect of increasing N.

In practice (i) is small ... By contrast (ii) and (iii) are very important, in particular (ii), which has the effect of flattening the graph.

On the face of it, the Ryle and Pooley results would seem to indicate an evolutionary universe; that is one in which sources are very distant and therefore viewed at an early stage of their development and have changed in some way. This could be an inherent change in the brightness of the object, or an increase in the number of sources per unit volume at large distances. In either case, a big-bang model would seem to be favoured above a steady-state model. In the sample, however, there is a mixture of normal radio galaxies with quasi-stellar objects (QSOs).

QSOs are massive, extremely small objects with large radio and optical emission. They tend to have large red-shifts, which would normally imply large distances. Their particular interest to us here is that a sample of QSOs studied in 1966 was found to have a steep $\log N - \log S$, plot whereas normal galaxies have a slope of -1.5. This is not unexpected, since the radio galaxies tend to be local, with small red-shifts. Objects with larger red-shifts could not all be allocated unambiguously as radio galaxies or QSOs. Where an optical identification was possible, the QSOs were found to have a steep distribution, as in Ryle and Pooley's work. Overall, the balance from radio source counts at the end of the 1960s seemed in favour of an evolutionary universe.

In 1965, Penzias and Wilson, working at the Bell Telephone Laboratories, discovered the microwave black-body radiation, a fossil relic from the early universe. Before considering this we will outline the 'standard model' of the universe put forward on the basis of recent developments in elementary particle physics.

10.4. History of the big bang

While obviously there are many uncertainties, the physics of elementary particles has developed to the point where a plausible scenario for the very early stages of the universe may be written. It must not be regarded as definitive, but some authors have considerable faith in it. The earliest time at which a statement can be made is the Planck time—10^{-43} seconds after the origin. The origin itself is inaccessible, because we do not have a quantum theory of gravity, and gravity is overwhelmingly dominant at this time. This time is about one billionth billionth billionth billionth billionth of a second. General relativity predicts that at the origin itself there will be a singularity, that is, an infinitely high density and an infinitely small size, but we cannot at present go before the Planck time, though an adequate theory may soon be developed. In the time before the Planck time an important particle may be the magnetic monopole, an entity first put forward by Dirac in 1932. Unlike present-day magnets these particles have isolated poles, not pairs of poles. They have never been observed with any degree of confidence. They would only be produced at very high temperatures or very high energies. The temperature at the Planck time is 10^{32} K.

The next interesting point occurs at 10^{-35} s. There are a number of problems at about this time, for example the sizes seem to be wrong. Looking back from the present size we expect that the diameter of the universe should then have been about a millimetre. Looking forward from the Planck time, however, it can only have been a billionth of a billionth of a billionth of this. Also the universe is rough and crinkly, whereas later on it must have been quite smooth. The curvature of the universe is wrong. Guth suggested in 1980 that there could have been a sudden expansion at this stage by an enormous factor, so just as on the surface of an expanding balloon, irregularities will have been smoothed out and the surface would have become flatter on the large scale. There were problems with this model, and an earlier time for inflation was tried. The theories are still controversial, but a majority of cosmologists probably believe in them.

As we will see, we cannot use light or any other electromagnetic radiation to observe the universe for the first million years, but there are other possibilities. Magnetic monopoles are one, of course, but so far the search for them has been unsuccessful. A class of particle theories called supersymmetric theories predict a series of partners for the familiar particles. The photino is the partner of the photon but much more massive. It should be produced in the early universe, but its detection would be very difficult. Gravitons and cosmic strings are other possibilities.

More promising is the primordial black hole. This would have been produced if the universe were very chaotic at about 10^{-28} s. At our present epoch, black holes will only be formed if the initial star is greater than the mass of the Sun, but at these primordial stages the pressure and temperature are sufficient to produce a whole spectrum of small black holes. In 1972 Bekenstein suggested that a black hole will have entropy, which means that it will have temperature. Hawking pointed out that if it has temperature it must radiate. This was contrary to the accepted picture of a black hole, which was thought to be sealed off completely. Hawking and his collaborators developed the theory of black-hole evaporation in the 1970s. If the mass is comparable to that of the Sun, evaporation effects will be negligible. If the black hole is small, say 10^{12} g or less, it will evaporate quickly and no primordial black holes will now be left. If it has the mass of a large mountain, however, about 10^{15} g, it will just have a lifetime comparable to the age of the universe. To detect it one should look for a burst of gamma-rays. Such bursts have been seen, but it is not clear whether they are coming from primordial black holes. At present all that one can say is that the early universe does not seem to have been very chaotic. This is a very active research field, however, and there will probably be further developments in it.

Returning to the early universe. At a millionth of a second it is showing signs of aging. The peculiar objects which infested space are beginning to disappear, or to freeze out.

At 100 seconds helium and deuterium are formed. These are important markers in the formation of the universe, and make up with hydrogen most of the universal matter today. The universe is opaque because of interactions. After 100 seconds little happened for 700 000 years, except expansion. There were a billion photons to each nuclear particle, but because the photon energies declined, whereas the rest mass of the nuclei remained constant, matter began to dominate over radiation. At the end of this period the universe became transparent and the cosmic background radiation appeared, which cooled as the universe expanded.

10.4.1. The cosmic microwave background

Of the fossil relics left over from the early universe, magnetic monopoles date back to the Planck time, primordial black holes to 10^{-28} s, whereas the cosmic background radiation goes back only to the somewhat geriatric age of 700 000 years. Unlike the other items, however, the background radiation is known to exist. The relic radiation was predicted by Dicke, before it was discovered by Penzias and Wilson. At the time the universe became transparent the temperature of the radiation was about 1000 K. It has now cooled to 2.7 K. The peak of the spectrum is now in the microwave region.

The best measurements of the background radiation have been obtained by the COBE satellites. The radiation is extremely uniform and isotropic, but an irregularity of 1 part in 10^5 has been found. This variation was sought because galaxy formation implies some small non-uniformity at early stages. The background radiation is generally considered to be an extremely strong piece of evidence in favour of the big-bang model and against the steady-state universe. Many former supporters of steady state deserted from the cause in 1965. Contopoulis and Kotsakis say [6]:

> It [steady-state theory] cannot account for the origin of the microwave background radiation. All attempts to attribute this radiation to galaxies or other sources in the Universe have failed. The only plausible explanation for the origin of the radiation remains the proposition that the universe passed through a state of very high density and temperature. This explanation is in complete contradiction with the 'perfect cosmological principle' and therefore with the theory of continuous creation.

10.4.2. Helium and deuterium production

Stars like the Sun contain about 25% of helium. In the universe as a whole there is one helium nucleus for every ten hydrogen nuclei. Theories of nucleosynthesis in stars predict 1–4% helium. Deuterium was detected by the Copernicus satellite in 1973 at one part in thirty thousand of hydrogen, again much larger than expected from stellar production. These elements are believed, therefore, to be primordial and generated in the big bang. The standard model predicts generation at about 100 s. Thus we have further evidence against the steady-state model. The deuterium has a further significance. It depends strongly on the density of the universe at 100 s, and is therefore a marker of this quantity. The density will ultimately decide whether the universe will go on expanding indefinitely, or whether the expansion will slow down, stop and reverse, contracting to a singularity.

10.4.3. The deceleration parameter q

Returning to figure 10.2, we see that the individual points are somewhat scattered, but tend not to favour $q = -1$, the value for the steady-state model.

Thus we have four pieces of evidence against a steady-state universe:

1. The deceleration parameter.
2. The radio source counts.
3. The cosmic microwave background.
4. The amount of helium in the universe.

In addition we have the close agreement in the age of the universe determined by radioactivity measurements and the Hubble constant. Also Einstein's equations predict an origin at a singularity. The evidence in favour of the big bang seemed overwhelming, and very few cosmologists accepted the steady-state model after 1965.

10.5. Steady-state fights back

The inflationary model is relevant to the argument, because the steady-state model is mathematically equivalent to it, with, of course, the limitation that inflation lasts only for a tiny fraction of a second, whereas the steady-state expansion goes on for ever. Barrow points out that the isotropy of the universe can be explained in the steady-state model but not in the big bang. This was realized in 1963 by Hoyle and Narlikar.

In 1980 Hoyle proposed a new version of the model. Creation of matter was concentrated in white holes, the converse of a black hole. Effectively there are many big bangs, each one generating its own universe. He made an interesting point: the evolution of life is so unlikely that it will take much longer than $1/H$, perhaps as long as 10^{21} years. There are great uncertainties in this, however. (An alternative view which he put forward was that life could have been brought to the Earth by a meteorite.)

At this stage, some doubt had been cast on the whole red-shift relation. Working with the 200 inch telescope on Mt Palomar, California, Arp [7] observed a number of strange associations. Luminous connections between galaxies with very different red-shifts were found. More striking still, QSO with large red-shifts were seen in apparent association with galaxies which had small red-shifts. There was some argument about the statisics of the effect. Arp concluded in 1979 that there were far more associations of this kind than would be expected statistically on a random basis. Burbidge [8] accepts this evidence for intrinsic red-shifts in some objects. Since three of the arguments for the big bang involve the red-shift, and since most of the objects observed at large distances are QSOs, this casts some doubt on the measurement of deceleration parameters. A fierce argument has raged about the statistical interpretation of Arp's observations, and this is not yet resolved. Thus the battle is perhaps not fully settled, but no one has as yet produced an explanation of the cosmic microwave radiation in terms of a steady-state model, and most cosmologists therefore accept the big bang model.

Chapter 11

MULTIPLE VERSUS PLURAL PARTICLE AND PHOTON PRODUCTION

The multiple/plural controversy is now largely forgotten, but it is worth recording, if only because it attracted the work of five Nobel Prizewinners (Heisenberg, Fermi, Blackett, Powell and Landau) as well as many other distinguished physicists. It had two phases, one in the 1930s, and one in the late 1940s and early 1950s. The first concerned the production of photons and electron pairs in fast electromagnetic interactions. The second concerned the production of mesons in nuclear collisions. Both of these controversies are now regarded as resolved.

Some of the arguments were carried out in print but mainly they were a matter of verbal conflict at conferences, preserved only in the memory of the participants. I was a very junior member of a plural group in the meson controversy; and took a deep, if silent, interest in the heated exchanges.

Plural production meant the production of particles one at a time but possibly in a sequence. Multiple production meant the simultaneous production of several particles (or photons) in a single interaction.

11.1. Cosmic rays

The idea of using fast nuclear particles as probes to examine the nucleus itself went back to the work of Rutherford and his collaborators in Manchester during the period 1908–21. It is an enormously successful technique which survives to the present day. With it Rutherford discovered the nuclear atom and also the existence of nuclear forces, which are stronger but more short-range than electrical forces. His projectiles were alpha particles from radioactive sources. These were limited in their effectiveness, because being positively charged (two protons in an alpha particle) they were repelled by the positive charge on the nuclei under examination. They were not energetic enough to overcome this electrical force and penetrate into the nucleus, though the occasional one showed by its behaviour that it had been affected by the nuclear forces.

Radioactive sources were inadequate, therefore, to probe deeper into the nucleus. More energetic and intense generators of particles were required. Accelerators were in their infancy in the 1920s, and were not to make significant contributions to the problems of elementary particles until the late 1940s. Therefore, physicists turned to the natural source of very high-energy particles, cosmic rays.

Hints of the existence of a high-energy natural radiation went back to Coulomb's work in the early nineteenth century. He found that charged electroscopes slowly lost their charge, no matter what efforts were made to insulate them. Part of this loss was in fact due to faulty insulation but there was a residual effect that defied all efforts to remove it. This was confirmed in a number of experiments at the end of the nineteenth century, using ionization chambers. The ionization chamber consists basically of a sealed vessel with two electrodes, filled with air or a noble gas, often at high pressure. The electric current between the electrodes (one of which is at a high voltage) indicates the ionization and hence the number of charged particles passing through the chamber each second. There are some complications, but basically this is how the system works.

Experiments in the years 1900–10 indicated that the ionization decreased if the chamber were lifted above the Earth's surface. Some rather inconclusive work was carried out at the top of the Eiffel Tower. In 1911, however, Victor Hess found, in a number of balloon flights, that after an initial fall, the intensity increased with altitude, up to the ceiling of the balloon. Modern experiments show that at very great altitudes the intensity falls again, but that the primary cosmic rays are coming in from space. At sea level there is a residue left, which makes a significant contribution to the total ionization. It can be divided into three components: the hard or penetrating component, the soft or electron–photon component and the neutron component. The penetrating component is dominant, and about five particles pass through the average human head every second. Occasionally a high-energy proton finds its way to sea level. We will return to these later.

It was realized in the 1920s that the sea-level components were secondary to the primary cosmic radiation, but neither the primary nor secondary components were identified at this stage. It was believed by the American physicist and Nobel Prizewinner Robert Millikan, that the primaries were very high-energy gamma rays, because of their great penetrating power. But if this were so, according to Dirac's theory of quantum interactions, the gamma rays should form electron–positron pairs and dissipate their energy. Also, the Dutch physicist Clay, and later Rossi, showed that the intensity of the sea-level radiation varied with latitude, being weakest at the geomagnetic equator. This implied, as shown theoretically by the Belgian Lemaitre and the Mexican Vallarta, working together, that the primaries were electrically charged; and later work showed that the charge was predominantly positive. The radiation was interacting with the Earth's magnetic field on approach, high above the atmosphere. Thus the situation in about 1930 was quite confused. A H Compton, another Nobel Prizewinner, embarked on an ambitious programme to measure the intensity simultaneously at sites all over the world, over a period of years. Hoffmann, using an ionization chamber, reported sudden bursts of ionization. These were called Hoffmann–Stosse or soft showers. It is these with which we will be dealing.

11.2. Techniques

Clearly new techniques were needed to disentangle a complex situation. These were forthcoming. The Geiger counter was an important device. It responded to charged particles passing through it. Typically it would be cylindrical, half a metre in length and 5 cm in diameter. Its response was independent of the amount of ionization released, above a threshold value. To make ionization measurements, a similar device, called a proportional counter, was used in a different mode.

More spectacular, though no more important, was the Wilson cloud chamber. This was developed by C T R Wilson at the beginning of the century. It works on the principle that if a volume of air saturated with water vapour is suddenly expanded by a limited amount, the droplets of water will condense preferentially on the ionized tracks left by charged particles passing through the chamber. The sensitive time is short, and after an exposure a complicated compression cycle is necessary. Cloud chambers, however, have been used not only at sea level and up mountains but also carried by large balloons to great altitudes. Effectively a picture of a track or more complex interaction is formed in droplets (figure 11.1).

The rich array of phenomena observed in the 1930s and 1940s was difficult to disentangle. Much attention focused on the soft showers and the positive electron or positron, predicted by Dirac and discovered in a cloud chamber by Anderson. The penetrating or 'hard' component also attracted interest in the early 1930s. A later technique, introduced by Blau and Wambacher in 1937, was the photographic emulsion. It did not reach its height until it was taken up by the Bristol group under Powell in collaboration with Ilford Ltd in the late 1940s. Photographic emulsions were developed which were far thicker and had far more silver than ordinary film. They were stacked together with no intervening glass or celluloid; then, after exposure, usually at high altitudes, individual films could be taken and mounted on glass plates for development handling and analysis. With this technique, nuclear interactions and their products could be followed through several plates and studied in detail. With this

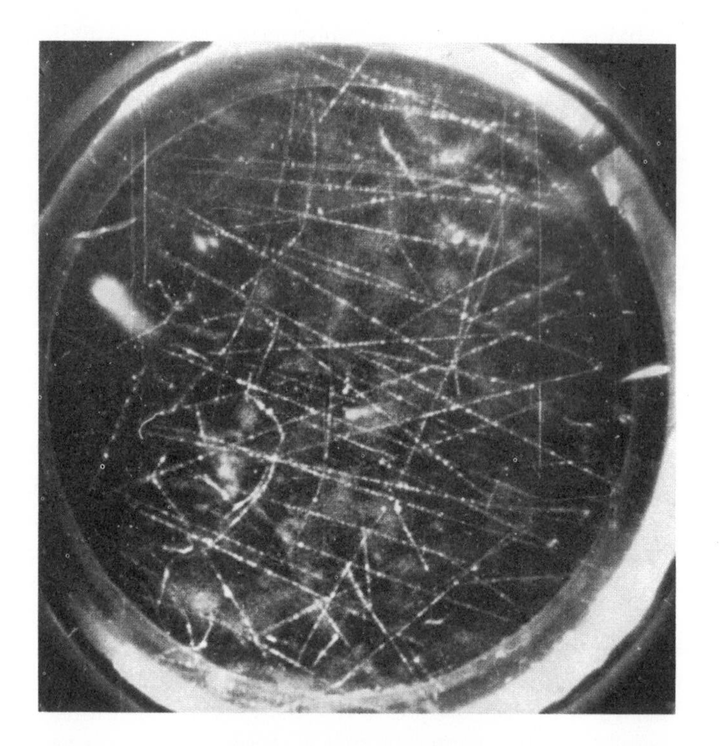

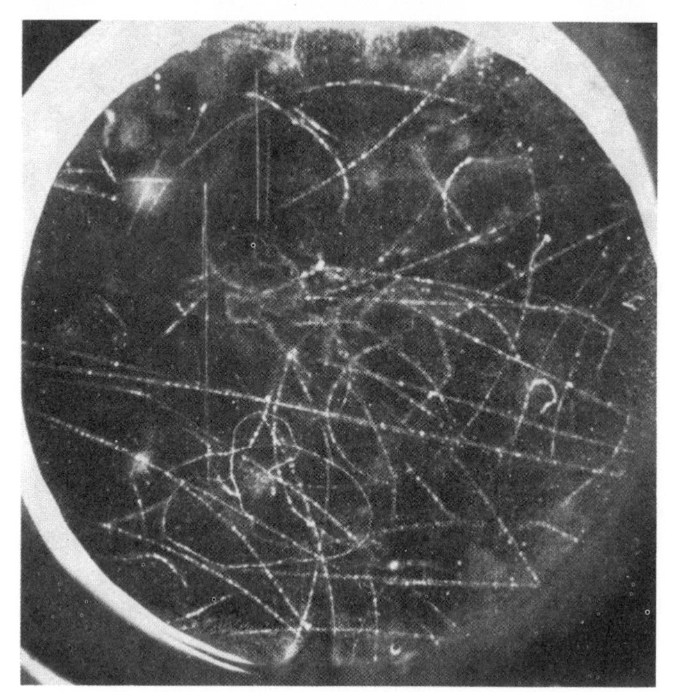

Figure 11.1. *Particle shower in a cloud chamber. Courtesy of Sir Bernard Lovell.*

technique, Bristol led the world in elementary particle physics for several years. Rochester (New York State) also developed a technique called the emulsion chamber, in which emulsions were sandwiched between thin plates of lead or other materials. These assemblies, called emulsion chambers, were suitable for observing very high-energy interactions, and were carried to great altitudes by balloons.

11.3. Soft showers

The Hoffmann–Stosse were the first examples seen of the soft shower, but the nature of the phenomenon became clearer with the work of Blackett and Occhialini in 1934. Using a cloud chamber with a magnetic field, they found a large burst of particles, subsequently identified as electrons, entering the chamber. Other examples originated in the lead plate inside the chamber (figure 11.2). Now this would not be expected in the accepted theory of electron interactions.

Quantum electrodynamics, as the theory was called, was the creation of Dirac, Heitler and others. It stemmed from quantum mechanics and developed in the early 1930s. An important constant in the theory is called the fine structure constant. It has the numerical value 1/137. The multiple/plural controversies hinged, to a great extent, on the value of this constant and of the corresponding quantity in the nuclear force field.

In the soft showers, two processes are important: bremsstrahlung (braking radiation) and pair production of electrons and positrons. A third process, less important at high energies, is ionization, i.e. the ejection of electrons from the atomic shells. Bremsstrahlung is the emission of a photon when a high-energy electron is deflected in the electric field of a nucleus. Pair production was predicted by Dirac and, as is indicated by the name, is the production of a positive and a negative electron in the field of a nucleus, by a photon. In both these processes the probability of n particles being produced is $(1/137)^n$. For example, the chance of two photons being produced in a bremsstrahlung event is 1/137 times smaller than that of one being produced. Hence multiple production is unlikely. (In pair production two or some other even number of particles must be produced to satisfy conservation of charge, but the chance of four or six being emitted is very small.)

The workers on soft showers, in both cloud chambers and Geiger counter arrays, were faced with an apparent breakdown of quantum electrodynamics. Heisenberg suggested that a modification of quantum mechanics, called a non-linear theory, would predict multiple production of photons in bremsstrahlung. These would in turn produce electron–positron pairs. Many theoreticians were reluctant to take such a step, and looked for alternatives. Bhabha and Heitler, and Carlson and Oppenheimer, independently developed the theory of an electron–photon cascade in 1936. In a given material, the average distance traversed by an electron before producing a bremsstrahlung photon is about the same as the average distance travelled by a photon before producing an electron–positron pair. This distance is known as the radiation length. In air it is about 300 metres but in lead it is only half a centimetre.

Thus, if we start with an electron, it will travel on average one radiation length before producing a bremsstrahlung photon, which will typically have one half of the electron energy. The photon will travel on average a radiation length before producing an electron–positron pair. Meanwhile the original electron will have produced another photon. So after two radiation lengths we have two electrons, one positron and one photon, a total of four particles (counting photons as particles). After t radiation lengths we will have 2^t particles.

This process will continue, without multiple production of photons, the numbers of particles increasing rapidly, until the energy of the individual particles is insufficient for further cascading. The depth of this maximum in number occurs at the same point in all materials if the depth is measured in radiation lengths. At the depth of the maximum, ionization losses deplete the electron component and the photon energies become too small for pair production. We emphasize again that there is no multiple production in this process, except for the dual production of electron pairs, necessitated by the conservation of charge. All photons are produced plurally.

A cloud chamber with a lead plate 5 cm in thickness mounted inside it will have ten radiation lengths of lead for the cascade to develop. If the maximum were at the lower surface of the lead, a sufficiently energetic electron could produce a shower of about a thousand particles, apparently diverging from a small volume close to the base of the plate. It is not surprising that many workers believed in multiple production, though Heisenberg withdrew graciously and conceded that the cascade model explained the phenomena, without the need to invoke photon multiplicities greater than predicted by the bremsstrahlung theory. Round one, therefore, went to the pluralists.

The full cascade theory, in which fluctuations in position and the energy of interactions are taken into account is difficult, but the elegance of the mathematical methods required attracted a number of distinguished physicists in the 1930s. The calculation of the lateral spread of the shower proved particularly intractable, and was not satisfactorily tackled until the growth of simulation models in the 1960s, though approximate theories were available in the early 1940s.

Interest in cascades grew at this time with the discovery of the Auger or extensive air showers. These are giant cascades in the atmosphere, starting at a height of about fifteen kilometres and blanketing up to a square kilometre on the ground with fast particles. They are a mixture of soft and nuclear cascades, caused by the arrival of an extremely energetic primary. We will not consider these in detail.

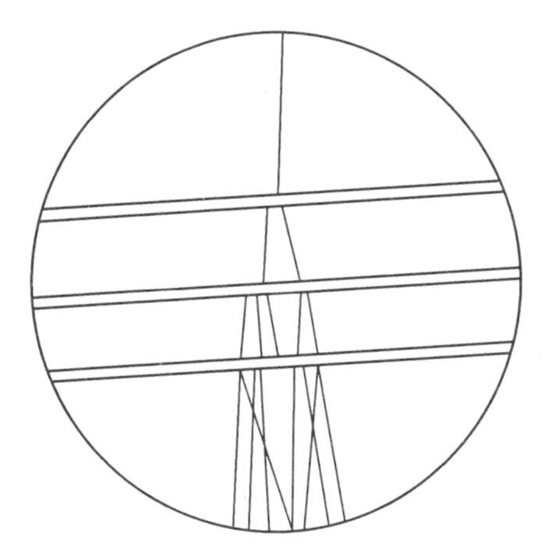

Figure 11.2. *Electron–photon cascade in a multiplate cloud chamber.*

The success of cascade theory does not prove conclusively that multiple production is not occurring. A relatively simple test is afforded by mounting a number of separated thin plates in a cloud chamber (figure 11.2). The development of the cascade is clearly seen, proving that the generation of particles is not a multiple act, but a plural cascade in which individual interactions of electrons are single and the cascade builds up slowly.

The lateral spread of the shower depends on the material. In lead, as we have seen, it is small, so that shower particles are projected back to a point near the base of the lead plate in the cloud chamber. A few showers, however, are different. Not only do the tracks project back to a point in the interior of the plate, but the behaviour of the secondary particles is quite different from that of electrons, penetrating centimetres of lead without cascade multiplication. These are called penetrating showers.

11.4. The cosmic ray beam in the atmosphere

It is now known that the cosmic rays striking the atmosphere are almost entirely nuclear in character, about 90% of the particles being protons, and most of the remainder being helium and heavier nuclei. There is a very small contribution from electrons and gamma rays. The primary particles interact, typically at an altitude of 15–30 km, with nuclei of oxygen or nitrogen in the air. The results of these collisions are the ejection of nucleons (neutrons or protons) and the production of mesons. Antiprotons may also be produced in small numbers. Mesons are particles with mass intermediate between electrons and protons. We will consider them later.

After the first interaction, the secondary particles move downwards, increasing in number as they cascade through the atmosphere. The number of particles reaches a maximum at about 15 km in altitude, and declines thereafter towards sea level. The composition of the beam changes with altitude. At 15 km there are a large number of energetic nucleons and nuclear active mesons, with electrons and photons. At sea level the majority of particles consist of inert mesons, called muons, which interact only by ionization and other electromagnetic processes, and can therefore penetrate to great depths in the ground. About a third of the particles are electrons and photons. There is an occasional high-energy proton and a greater number of low-energy neutrons. It has recently been suggested that cosmic rays contribute to global warming.

11.5. The meson

The problem of electric forces between electrons was solved by the stratagem of assigning to the photon the function of an exchange particle. When the strong but short-range nuclear force was discovered, it was realized that photons would not suffice. The young Japanese physicist Hideki Yukawa suggested that if a nuclear-active particle of mass about 200 times the electron mass, roughly one tenth of a proton, existed then the nuclear force would be explained. His idea was not immediately accepted, largely because it was dismissed by Niels Bohr, then on a visit to Japan. Bohr said that following the success of electron–photon cascade theory, all cosmic-ray phenomena could be understood on the basis of current quantum theory. It was a rare but serious mistake for Bohr to make, and one which was very distressing for Yukawa.

The meson was discovered by Neddermeyer and Anderson in 1937, narrowly beating Blackett's group by a few weeks. Measurements of ionization and momentum in a cloud chamber with a magnetic field gave a mass of about 200 electron masses, as Yukawa had predicted. In quite a short time, however, problems emerged. If this were the Yukawa meson, it should interact strongly with nuclei in the air or ground, but it did not, and lost energy only by ionization. The problem was solved in the late 1940s, using the nuclear emulsion technique, by the Bristol group. The penetrating meson in cosmic rays was found to be formed by the decay of a shorter-lived nuclear-active meson, which was identified with the Yukawa meson. It was called the pion, and existed in three forms: positive, negative and neutral. The charged forms lived for about one hundred-millionth of a second, decaying to the inert meson, called the muon, which lived for two millionths of a second. The neutral pion decayed in a very short time to two gamma rays, so did not have time to interact with nuclei. There was no neutral muon.

11.6. The theories of meson production

After the pion was discovered, great interest was aroused in the process by which it was produced. A theory had already been constructed by Hamilton, Heitler and Peng (HHP) in Dublin in the years 1942–5. Since Heitler's work before the war had mainly been on such processes as bremsstrahlung, it was natural that the new meson theory should proceed by analogy with these electromagnetic interactions, which had been so successful in the plural theory of the soft shower. In the HHP theory the cross-section for interaction of a proton with a proton or neutron to produce a meson or mesons, followed a geometrical model. It was assumed that only one meson was produced in a given nucleon–nucleon collision, and the collision was highly elastic, that is, the primary particle retained a high proportion of its energy.

In the late 1940s, Heitler and Janossy in Dublin developed a model of plural meson production. At about the same time, Lewis, Oppenheimer and Wouthuysen, Fermi, Heisenberg and Landau also developed theories. The Heitler–Janossy model is best considered separately from the others, since it was totally different in form. In a 1949 paper they consider the 'stars' seen in nuclear emulsions [1]:

> In recent papers by Brown, Camerini, Fowler, Heitler, King and Powell the relative frequency of stars with a given number *n* of thin tracks is measured. There can be little doubt that these events in the photographic emulsion are identical with the penetrating showers previously found and observed in counter and cloud chamber experiments by Janossy, Rochester and others. If this is true, then it is also highly probable that we have to deal here with the process of meson production by fast nucleons (and indeed, roughly half of the stars are initiated by charged, half by neutral particles) and that at least a considerable fraction of the thin tracks are mesons. Some of the tracks are also likely to be fast protons.
>
> For the interpretation of these events, two widely different hypotheses have been put forward:
>
> (i) The occurrence of groups of particles is explained in a natural way as due to a multiple collision of the primary nucleon with several of the closely packed nucleons in a compound nucleus, O, C, N or Ag, Br in the photographic plate. We call this process, following J G Wilson, plural production.
>
> (ii) The groups are assigned to a single elementary process in which the fast nucleon would produce several mesons in one single collision with a single nuclear nucleon (genuine multiple process). It is assumed here that the emission of several mesons in one act is more probable than that of a single meson. The process (ii) would be a type of process for which so far no analogue is known in physics, in so far as the emission of the minimum number of particles or quanta (compatible with conservation laws) is always the most probable event. A decision between the two interpretations therefore involves a rather fundamental issue and

> would be of great importance for the future development of meson theory. In this paper we examine the question whether the data available to date can be understood by the assumption (i) of pure plural production. It will be seen that this is the case...

A few words on this passage may be helpful. The geometry of the interaction is quite different from that in bremsstrahlung. There the chance of a second interaction in an atom is negligible, since the nucleons in the nucleus act collectively in generating the electric field. In meson production, the nucleons in a nucleus act independently but are closely clustered. A simple calculation shows that the chance of a second or third collision within the nucleus is quite high. Thus in meson production, plural processes will be quite likely, provided the incident nucleon retains a fairly high proportion of its initial energy at each collision. Also, recoil nucleons may themselves produce mesons.

The idea that a plural process of this kind could produce up to fifty fast mesons and nucleons sounded strange at the time. Nevertheless calculations by Heitler and Janossy, and later by Messel and his collaborators, found that such showers could be created in a plural mechanism. There is, however, a fundamental problem about this model. We have seen that the probability of producing n particles in a bremsstrahlung interaction is proportional to $(1/137)^n$ where $(1/137)$ is the fine structure constant for electromagnetic interactions. In the nuclear field, however, the corresponding constant is much larger, perhaps as high as 1 or even greater. Multiple production of mesons, therefore, would seem quite likely in a single nucleon–nucleon collision. This assumes, of course, that the bremsstrahlung analogy is valid at all. Many theoreticians thought it was not.

Nevertheless, the hypothesis of plural production was pursued, and almost elevated to the level of religious dogma by some pluralists. They showed, in an ingenious and long-lived controversy, that experimental results hitherto interpreted as demonstrating multiple production, could in fact be explained on the plural model and, moreover, that some results could not be understood on a multiple model at all. There was, however, one piece of evidence which seemed very difficult to explain on a plural model. Most 'jets' in nuclear emulsion had a number of short heavily ionizing prongs from the origin. These were taken to indicate the presence of a heavy nucleus in the interaction (figure 11.3). Occasionally, however, there were stars which had no heavy prongs. These were taken to indicate a pure nucleon–nucleon collision with multiple production.

11.7. Multiple meson production theories

Lewis, Oppenheimer and Wouthuysen in 1948, working in Princeton, produced a theory of multiple meson production. Again, it was based on an analogy from another branch of physics, in this case from atomic

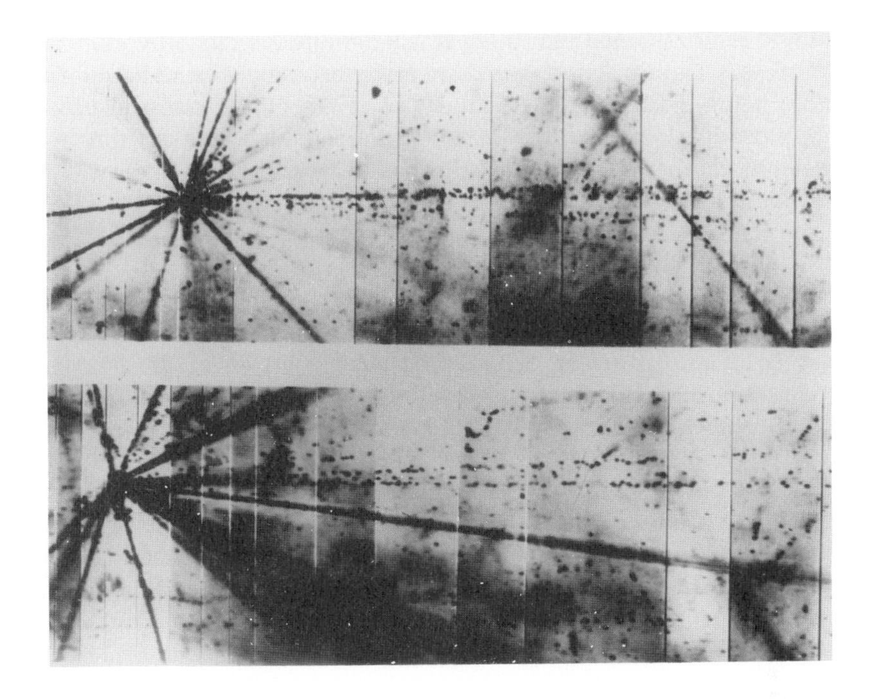

Figure 11.3. *Jets produced by an accelerator in nuclear photographic emulsion. Courtesy of A C Breslin.*

physics. The interactions were assumed to be quite weak. For a proton–proton collision, they found meson multiplicities of 5 at a primary energy of 10 GeV, and 9.1 at a primary energy of 30 GeV. These numbers were very high compared with plural theory.

11.7.1. Hydrodynamic theories

Heisenberg in 1949, working in Göttingen, used the analogy of turbulence in a viscous fluid to treat nucleon collisions. The dissipation of energy in mesons corresponded to the production of eddies. It was assumed that the eddy production was terminated by meson emission before equilibrium was reached. He found that the number of mesons produced varied roughly as the square root of the primary energy. This agreed with work by Wataghin in 1948. The number of mesons produced at 10 GeV was 12.6 and at 30 GeV, 20.4. In 1952, he produced a modified theory in which a more formal model of the meson production was used. At 10 GeV the number of mesons was predicted to be 3.6, far less than in the previous treatment. Landau extended this theory [2].

11.7.2. Thermodynamic theories

In a paper published in 1950, Fermi in Chicago outlined a theory of meson production which lies at the opposite extreme from that of

Lewis *et al.* It assumed that there was a definite interaction volume, and that mesons and nucleons shared energy, reaching a state of equilibrium, before being ejected. The treatment used essentially classical physics, with close analogies with the theories of radiation worked out in the nineteenth century. It was presented as a very tentative theory, which was Fermi's style. Later he produced a high-energy theory.

11.8. Experimental work

Many experiments were done with Geiger counter assemblies, cloud chambers and emulsions between 1948 and 1953. All of these used cosmic-ray nuclear particles because energies of particles from accelerators were insufficient. Generally the results appeared to favour multiple production, but this could always be explained away by the pluralists. Violent battles raged at conferences.

Finally, an accelerator, the Cosmotron at Brookhaven in New York, was able to produce sufficiently energetic particles. Multiple production was seen unambiguously. There was some uncertainty because there was some instrumental bias against the detection of plural events, but by and large the controversy was ended and today a modification of Fermi theory is used. The battle is over.

Chapter 12

MISSING MAGNETIC MONOPOLES

This chapter describes not so much a conflict as a mystery, as yet unsolved. It is about an observation which, if true, was of great importance, but which was never observed again. It happened in Stanford University, California on 14 February 1982.

In Maxwell's work on electromagnetism, there was a remarkable parallel between electric and magnetic fields, but with an exception. Electricity had clearly defined charges. The magnetic field on the other hand did not seem to have any magnetic charges associated with it. All the magnetic effects observed arose from circulating electric currents, or other movement of electric charge.

When constructing his fundamental equations, therefore, Maxwell deliberately left out any terms relating to magnetic charges or magnetic currents. If such effects were discovered, terms could easily be put in to describe them. So the matter rested for fifty years. No evidence was found to show that magnetic charge existed. There were, however, such things as bar magnets, which appear to have a north pole at one end and a south pole at the other. This seems to demand magnetic charges of both kinds. But if a bar magnet is cut in two one does not obtain two separate charges, but two smaller bar magnets, each with a north and a south pole. It is not possible to separate magnetic poles in this way; they always turn out to be a consequence of electric currents, and a north pole is always associated with a south pole.

In 1931, however P A M Dirac pointed out an anomaly that had largely been overlooked. Quantum mechanics had been enormously successful, but one of the things which it did not explain was the uniformity of electric charge. Why do all electrons have the same charge, and why is it equal and opposite to the proton and positron charges? He showed that if we assume the existence of an isolated magnetic charge, then an equation can be set up which predicts both the electric and magnetic charges and the relation between them [1]. Given the measured value of the electron charge, that for the magnetic charge can be calculated. The new entity was called a magnetic monopole. Even if only one of them existed in the whole universe the Dirac theory would apply. The mass of the monopole was not predicted by the theory and there were no other clues to it.

The suggestion raised a certain amount of interest because of the great respect in which Dirac was held, but there was no serious attempt to find the monopole for some years. In 1947 Dirac spent a sabbatical in Princeton and worked again on the monopole. He produced much more detailed predictions [2]. The value of the magnetic charge was found to be 137/2 times that of the electron, measured in appropriate units. Since processes such as ionization tend to go as the square of the charge, they would be about 4000 times as strong as for the electron. Thus a fast monopole in emulsion would produce a very heavy track rather like a totally ionized heavy nucleus. Another striking effect is the acceleration in magnetic fields. Unlike electrons, which spiral in circles, monopoles will be accelerated along the line of the field and will gain energy very quickly.

This property of direct acceleration was used by Malkus [3] at Chicago in 1951 to look for monopoles. An evacuated pipe, about two metres long, with an axial magnetic field, was pointed along the Earth's magnetic lines of force. At the lower end of the pipe was a nuclear emulsion. If a monopole entered at the top it would be accelerated down the pipe and produce a track in the nuclear emulsion. No events were seen but upper limits were set.

In 1960, I suggested [4] that the highest energy cosmic rays might be monopoles, which would be rapidly accelerated in the galactic magnetic fields. This was taken up by Alvarez and others, and during the 1960s a number of attempts were made to tear monopoles out of magnetic rock or sediments where they would be trapped over millions of years. Again, only upper limits were seen, and interest lapsed somewhat in the early 1970s.

In 1975, however, a new class of theories was developed, with the aim of unifying the strong nuclear force with electromagnetism and the weak force. These were called grand unification theories or GUTs. They predict very large numbers of magnetic monopoles, each with a very large mass; so large that they could not be produced in any accelerator envisaged today or even in high-energy cosmic rays. Only at the origin of the universe, close to the Planck time, would enough energy be present. Monopoles could thus be fossil relics of the big bang

The GUTs predicted that the monopoles would act as catalysts for the decay of protons, in addition to a spontaneous decay with a lifetime of about 10^{31} years [5]. This may seem a very difficult prediction to verify, but it is in fact perfectly possible in principle. A cubic centimetre of water has 10^{24} protons in it. A 10 m cube will have 10^9 cubic centimetres and 10^{33} protons. It will therefore have about 100 decays a year. By installing a large tank of water in a deep mine (to cut down the cosmic-ray background) and covering the walls with large phototubes to detect the Cerenkov light produced by the decay particles, the predicted rates can be tested. So far decays have not been seen and there has been no sign of catalysis by monopoles. The theories have some margin of reserve, however, so a final conclusion is not possible. Current predictions of the lifetime of the proton are as high as 10^{34} years, a number very difficult to test experimentally.

A somewhat different system has been used in an experiment in the Kolar Gold Mines in India. An assembly of proportional counters was used to plot the tracks of particles which apparently originated in the material. Several candidate events were seen, but they were very difficult to separate from cosmic-ray induced interactions.

GUTs present a number of problems, the most important from our point of view being the large number of monopoles produced. These monopoles will not be destroyed except by annihilation, which will be rare. Their concentration, however, will be drastically reduced by inflation to a tolerably low level. There are stringent limits on the number of monopoles which can be tolerated. We know from a variety of observations that there is a magnetic field in our Galaxy. If the density of monopoles were high, they would short-circuit and destroy the magnetic field. From the fact that it continues to exist, a limit, known as the Parker limit, can be set on the number of monopoles in the Galaxy.

Following the GUT predictions, a number of experiments were carried out to look for monopoles. Professor Blas Cabrera of Stanford University in California built a system involving a very sensitive detector of magnetic fields called a SQUID (superconducting quantum interferometer) which could respond to a single monopole [6]. On 14 February 1982, St Valentine's day, after 150 days of running, a single event was seen which had all the properties of a monopole. If the apparent rate (though of course with a single event it is not really possible to quote a rate) was a good measure of the true rate, this would imply a number greater than the Parker limit. Observations were continued but no further events were seen. A larger, more sensitive system was built and saw nothing.

The origin of the single 'St Valentine's Day' event has roused eager debate. Could it be a very large fluctuation in a genuine rate? It has been suggested that it was an elaborate hoax. If so it was very sophisticated and the perpetrator has not been found.

Underground detectors, designed for studying proton decay or neutrino astronomy, would also have been sensitive to catalysed decay, but no such decay has been seen. The Cabrera event remains the only one which fulfils all the requirements for acceptance as a monopole. If we exclude a hoax, the remaining explanations are rather few. Essentially the experiment consisted of a coil immersed in liquid helium at a very low temperature and shielded from the Earth's magnetic field. An enormous magnetic storm might produce such an event but that would surely be observed elsewhere. To the author the most likely explanation seems to be a partial component failure somewhere near the output, which just happened by bad luck to be of the correct magnitude.

A positive observation of a monopole would be of great importance in the theory of fundamental forces. It would also tidy up a number of less important problems, such as the acceleration of very high-energy cosmic-ray particles. This has been a mystery for the past fifty years. A monopole would be accelerated linearly along a magnetic field line and would rapidly reach energies of 10^{20} eV.

We have not yet considered the mass of the original Dirac monopole. There was in fact no indication of this in the theory. There is now some suggestion of a mass in the region of 10^{15} eV, and recently and more speculatively, zero mass. The charge is the same in all these theories, 137/2 times the electron value. This means that electromagnetic interactions such as ionization will be much stronger, though if the mass is large, bremsstrahlung will be reduced. Another development, supersymmetry, predicts even higher masses than GUT, 10^{28} eV or 10^{-5} g.

Altogether the monopole is something of a mystery. Ever since it was first put forward nearly seventy years ago, it has refused to go away, but pops up in theories over and over again in different forms. The only direct evidence for it remains Cabrera's event, and we are losing hope of a repetition of that. If it does exist, it will probably be found in the context of a high-energy situation, but we know from the Parker limit that there cannot be too many monopoles in the Galaxy to look for. Hope springs eternal, for it would solve many problems, but perhaps we should now be looking for reasons why it should *not* exist.

Chapter 13

CONCLUSIONS

Most forms of conflict are in some way regulated. The concept of a limited war is an example, and war itself requires, supposedly, disciplined troops and high technology. For those who wish to mediate in warfare there are the Geneva conventions, international treaties like the nuclear test ban treaty and if all else fails international war-crimes tribunals. In marital conflict there are guidance counsellors; if things get worse, mediators and if the worst comes to the worst divorce courts. Crime is still unregulated on the criminal side but the resources of the law and the judiciary use all the techniques available. The stock market is highly regulated in an attempt to avoid conflict and for the same reason industrial production is subject to innumerable health and safety regulations.

By comparison, scientific communication, dialogue and conflict seem to exist in a state of chaos largely unchanged since the seventeenth century. Anyone who has initial capital can set up a scientific journal with themselves as editor if they wish. They must persuade some people preferably but not necessarily reputable, to act as associate editors or referees, usually without pay. University libraries more or less have to subscribe (though this seems to be breaking down with shortages of library funds) if they want to remain leading libraries. Contributors are not paid; usually they pay the editor. Subscriptions to leading journals are in the region of £1500. There is no mediation (science has this in common with the humanities) between author and editor. This is the setting in which the greatest intellectual activity in human history operates. It operates in a partial vacuum, since a large proportion of the population does not understand what the scientist is doing. The system succeeds because one must publish to gain promotion, and somehow the anarchy is welded by the scientific community into a coherent whole, though it would be better to refer to communities, since most scientists only understand a small part of their subject. Kuhn estimates that a typical community would have about a hundred members. Since physics alone has about a quarter of a million people worldwide, it is clear that the participation is highly fragmented.

It is indeed astonishing that, by and large, this shambles seems to work so well. Its limit is the amount of money that libraries are prepared to spend on journals (and more important, the amount of money that governments are prepared to spend on accelerators, satellites, radio telescopes and other expensive items). Not all journals succeed; they do not usually stop publishing, but a general consensus arises amongst a community as to which are the prestigious journals and those most widely read. The arrival of electronic literature is unlikely to make things much better. Occasionally a journal makes a serious error. Bose was turned down by *Philosophical Magazine* and only rescued by Einstein. Einstein himself was fortunate that the editor of *Annalen der Physik* recognized the importance of the very strange papers submitted to him, and published them. Probably some papers are published which should not be, and, less frequently, good papers are turned down. There is a self-correcting process: scientists soon learn to avoid journals in which the refereeing or editing is defective, and a hierarchy of merit is established for both journals and papers. By and large, the best papers are published in the best and most widely read journals.

The system, if it can be dignified by that name, works well so long as there is agreement and dialogue. It breaks down when there is conflict, particularly when there is internal conflict between physicists. Journals do not usually like conflict, they want to publish verifiable results. A small number of journals publish comment but they are usually directed at a wide audience. Correspondence pages are confined to journals like *Nature* and *Science*, which fulfil a dual function. Review articles offer some opportunity for comment, but it is limited to much the same degree as an individual paper. The scope for controversy in print seems to be wider in the humanities.

The main battleground for scientific controversy is at conferences, where strong men may be reduced to tears by the response to their papers. Fortunately or unfortunately, these verbal battles do not usually appear in the conference proceedings. What do appear are the rapporteur papers. Here there is scope for comment, but it tends to be one-sided, since the responses to the rapporteur papers usually do not appear in the proceedings. Thus most of the conflict is carried out by word of mouth at meetings or conferences, or by correspondence. Correspondence seems to be declining in favour of e-mail. The author has noticed a marked increase in dialogue and argument among his colleagues in a large collaboration since e-mail became popular, so much so that one wonders where there is time for science in the first place.

The following list expands on the various interactions.

1. *BARGAINING* is a feature in political, military, industrial, legal and marital conflict. It hardly figures at all in scientific conflict. This

is because it is usually impossible to detach and abandon part of a scientific theory or set of scientific observations without destroying the structure altogether. Extra-scientific bargains are not unknown. Galileo agreed, it seems, to make a bargain with the Inquisition. If he recanted abjectly enough, he would get a light or nominal sentence. He was cheated. The sentence of perpetual house arrest was as savage as the Grand Duke would tolerate, as an indispensable ally of the Papacy, but this was not a scientific bargain.

2. *BLUFFING* has been treated extensively by Schelling in his *'Strategy of Conflict'*. It is obviously important in military conflict, but, while not entirely unheard of, it is rare in science. This is partly because one can seldom get away with it. Most scientific results can be checked by others. It occurs occasionally in dialogue at conferences.
3. *LYING AND FRAUD* are remarkably rare. Most scientists will not encounter either in their career. The penalties, if found out, are great. Also most scientists have a genuine commitment to their subject, which is not easy to pursue but requires hard continual effort. 'Smoothing' and selection of data are not, however, unknown, but they hardly amount to fraud.
4. *MEDIATION*, common in military, political and marital disharmony, is rare in science, as indeed in all academic life. This is probably because of the same factors we have noted in relation to bargaining. Mediation involves a give and take kind of openness which is seldom possible in science. One cannot sacrifice half of one's theory; it would become meaningless. There have been attempts at mediation. Paul Ehrenfest, a close friend of both Einstein and Bohr, tried to resolve, or help to resolve, the conflict between them in the 1920s. He was inevitably unsuccessful and this contributed to his suicide in 1933.
5. *EXTERNAL JUDGEMENT* arises in two contexts. The editors of journals, and the referees they employ, have a very important function. Frequently they turn papers down, not because the facts are wrong, but because the material is not considered appropriate for the journal. A determined author, however, can usually find a journal which will publish his paper.
6. *PHILOSOPHICAL CONSTRAINTS*. We have seen that positivism figured in the Boltzmann case, but it has, in my view, been exaggerated. Mach was a positivist, but Kelvin certainly was not. Some primitive philosophical concepts are involved in all scientific activity, but most scientists seem to distrust and dislike philosophers, and in their opinion philosophy of science, if tolerable at all, is only fit for those who are too old to contribute to genuine science. This view was held by Schrödinger, although, in addition to his greatness as a physicist, he was accomplished in philosophy of science. Physicists particularly resist any attempt to make scientific judgments on philosophical criteria, though they are ambivalent

about quantum mechanics. The Boltzmann case is a rather rare example of a scientific conflict in which philosophical criteria were important.

7. *DISPUTES WITH AUTHORITY*. Here some judicial element is always present, but science may be totally absent. In the cases we have considered, some official and non-scientific tribunal gave a verdict. Hypatia, of course, was murdered without any formal judgment, but Bacon was judged by his religious order, Bruno and Galileo by the Inquisition and Oppenheimer was judged by a tribunal set up by the Atomic Energy Commission. None of these were scientific bodies though there was one chemist on the Oppenheimer tribunal. In all these, bargaining, bluffing, the adversarial process, and possibly lying were freely employed.

13.1. The effects of conflict

Is conflict harmful to the progress of science? We have seen that some philosophers of science, particularly Marxists, believe that conflict is not only beneficial but necessary for progress. It may be so but certainly it is at least uncomfortable for some of the participants. We make a distinction between conflict with authority, and internal conflict. The former is not only uncomfortable but life-threatening. Moreover, in the judicial enquiries we have discussed there was no movement forward; they were retrogressive. Indeed there were no scientific arguments even in the Oppenheimer case. It could hardly be held then that these disputes brought about any progress.

Internal conflicts are different. In the Einstein–Bohr conflict, the attention of many scientists was drawn to fundamental problems of quantum mechanics, articulated by scientists of genius. This would seem to be good, though it cost a life, but it did not in itself cause any movement forward. Einstein, Podolsky and Rosen put forward an important theorem, but it was not answered in detail by Bohr. The discussions only became fruitful thirty years afterwards, when recollected in tranquillity by John Bell. Is this delay a general feature of conflict? It would perhaps be more accurate to say that conflict causes a stagnant area or eddy in the flow of the river of science. This is temporary and may be by-passed. Thus Kepler by-passed Galileo on the heliocentric theory. Science moves around the wrestling protagonists and quietly goes forward without them. To illustrate this, what did the central figures do after the conflict was over?

13.2. After the conflict

After Galileo's ordeal with the Inquisition he produced his best book, but on a subject removed from the heliocentric model. In the mean time Kepler had produced the *Rudolphine Tables* (1627), which moved the model on to its next phase.

Hypatia and Bruno did not survive their conflicts, and Boltzmann also died before his conflict was resolved. Bacon was silenced. All these scientists were removed from their struggle. Oppenheimer went back to his job as Director of the Institute for Advanced Studies, but produced little physics. Einstein never entirely gave up his interest in the fundamentals of quantum mechanics, but put most of his time into the search for the unified field theory. Blackett, defeated over strategic bombing during the war, but vindicated by post-war surveys, went back to cosmic rays and developed an interest in palaeomagnetism. The multiple and plural supporters went on to other things, as did the steady-state supporters. The man who was totally destroyed by the collapse of his subject was Blondlot, to whom N-rays meant everything. By and large conflicts seem to be terminal, at least as far as a given topic is concerned.

13.3. Acceptance by the scientific community

The scientific community is a rather amorphous entity. There are about a quarter of a million physicists in the world, but we have seen that Kuhn considers that the active community in a particular issue is only about a hundred people. Since there may be collaborations, the number of independent parties may be as small as ten.

In spite of its disorganized character the physics community is effective. It reached its highest point in the 1920s and 1930s, when the physics population was small and most academic physicists knew each other personally. Copenhagen, Göttingen, Paris, Cambridge, Rome and Manchester were havens in Europe for wandering scholars. The growth in the number of physicists and the discovery of the strategic character of physics have diluted this, coinciding with the introduction of the word 'community' to describe it, but a strong sense of solidarity remains.

There is no formal procedure for the acceptance or rejection of a concept or a set of results. Publication in a refereed journal is a point in their favour, but by no means final. While the community is fairly democratic, there are senior figures whose judgment is respected. These are not necessarily the oldest members of the community. With rapidly changing techniques and concepts, physicists tend to peak in their thirties, and by the time they are fifty-five are largely buried in administration.

This is the procedure if a discovery or theory is not contentious. It has been analysed in detail by Kuhn and his co-workers. If there is conflict, though, the community does not cope so well. Publication of a paper is not enough to ensure its acceptance, and as we have seen there are no judicial, mediating or bargaining functions. The community tends to split, usually into a large majority and a small minority which includes the authors of the paper. Suspended judgment sets in while the community waits for independent proof or disproof. Usually while it is waiting there are strong opinions on the proposed discovery. In time, possibly as much as five or ten years, confirmation or disproof is found. The system adjusts to absorb the new piece of information and build it into the structure of physics. Sometimes, quite often in astronomy, the process does not work so well. If a flare or supernova explosion is observed by one group only, that exact observation cannot be confirmed by other observers. The effect may be seen in a later stage of its evolution and for this reason there is a complex international warning network which alerts astronomers quickly to transient phenomena of this kind. Nevertheless some very interesting phenomena cannot be confirmed and one can only hope for a repetition. Obviously co-ordinated observations by two or more groups are helpful.

There are mechanisms of approval and acceptance. Most learned societies and academies have medals and other awards, and ultimately there is the Nobel Prize. There is usually agreement in the community about the subject of the Prize, the Nobel Foundation is cautious in the extreme and avoids contentious subjects. Occasionally there is ill-feeling about the personnel who receive the award—a feeling that more people should have shared in it (though each prize is limited to three people). In recent years most Nobel Prizes in Physics have been given for high-powered and expensive experiments, so that small groups have been almost priced out.

Medals and prizes, therefore are given mainly for established and accepted work. They are not used for resolving conflicts. Where distinguished scientists are in contention, an academy will usually contrive to make awards to both, after a suitable interval. By and large the community is civilized in its behaviour, with occasional outbursts of savagery. A majority of its members have a strong and genuine commitment to their subject, and self-interest is probably less than in most professional groups. However, they have a passionate attachment to their own results. Thus there is no formal mechanism for resolving conflicts. The process is rather one of seepage, leading to general consensus and incorporation into the general body of knowledge and understanding. This is effective but slow—a hundred years for Copernicus, thirty years for Boltzmann, twenty years for Einstein. This seepage is not a passive process. Frequently there are intensive efforts to validate or disprove other physicists' results, but the process of acculturation is slow.

13.4. The attitude of the community

13.4.1. Institutional conflict

Here the individual is generally seen as a martyr by colleagues who believe him to be right. Galileo had many visitors at Arcetri after 1633, including John Milton and Thomas Hobbes. Though he was not allowed to leave his house, there were no restrictions on his visitors.

Oppenheimer attracted much support from physicists in the US after his security clearance was withdrawn, and he too was seen as a martyr. There was never any question of his losing his job; the main blow was to his self-esteem. It is not known how Bacon's Franciscan brethren viewed his incarceration, but during the late Middle Ages he became somewhat of a cult figure, with a reputation for magical powers. He was not, however, generally regarded as a martyr. Bruno became a martyr figure, but not until the late nineteenth century during the period of the first Italian monarchy. This was when the enormous statue of him was erected in the Campo di Fiori. At the time of his execution he was regarded as a heretic and aroused mixed feelings. The scientific side of his thought was hardly noticed. As a person, he was generally disliked.

13.4.2. Conflict between physicists

Most conflicts of this kind last over several years as experimental evidence accumulates. The defeated protagonist is not generally regarded as a martyr.

13.5. Things in common

We have identified a number of negative features held in common where there is conflict. There is no bargaining, bluffing or mediation. Are there positive features in common? The formal structures for those in conflict with institutions show some similarities. In 1958 de Santillana made comparisons between the Galileo and Oppenheimer cases [1]:

> It might be a mere game of tally, but it is tempting to establish a one-to-one correspondence between the actors of two dramas three centuries apart and to follow them through parallel convolutions. One might say, for instance: for Holy Office read 'AEC Board', for 'Caccini' read 'Crouch', for 'Lorini' read 'Borden', for 'SJ' read 'Strategic Air Command', for 'report of the preliminary Commission' read 'Gray–Morgan majority report', for 'Grienberger' read 'Teller', for 'Certain German mathematicians' read 'a certain Dr Malraux'.
>
> As for the hooded figure of Michaelangelo Seghizi de Lauda, the craftsman of iniquity, the number of highly placed contestants in public life and in the communications empire would make it invidious to choose.

The lives of Galileo and Oppenheimer share the same kind of time pattern, as noticed earlier: rise from early promotion to a position of eminence in early middle age, followed by sudden disaster and loss of standing. They were both non-torturable, as Graham Greene has put it, though there has been a spurious allegation that Galileo was shown the instruments of torture. The favour of the Grand Duke, an indispensable ally of the Papacy, as well as Galileo's age prevented any physical ill-treatment.

For those who hold the conspiracy theory of history, the question arises as to whether these persecutions were the results of plots. Professor de Santillana [2] puts forward an ingenious theory that Galileo's condemnation was directed not primarily against Galileo, but at the unfortunate Father Riccardi, 'Padre Mostro', who had given him his Roman imprimatur. In Oppenheimer's case it would be easy to put forward conspirators; he had enemies. But had they the power to initiate an enquiry of the kind carried out by the AEC? One man who had was General Nicholls, Manager of the AEC and former aide to General Groves. Could this be an indirect attack on General Groves, Oppenheimer's protector, from within the AEC? It seems possible but far-fetched, since Groves had now retired. Army Security is a more likely source. This could hardly be described as a plot.

Equally, the Pope's patronage of Roger Bacon could have caused jealousy and imprisonment as soon as the protector died, but he was already imprisoned before this happened. Certainly there was a conspiracy against Vavilov, which was successful. Boltzmann's persecution could be described as a conspiracy, but the term hardly seems accurate. In the case of Bruno, the remarkable thing is that he was not arrested earlier by Lutherans or Calvinists during his wanderings about Europe. Einstein may have been the subject of an anti-semitic campaign, but this was not responsible for the battle over quantum mechanics. No Nazi scientist played a significant part in it.

So, by and large the conspiracy theory does not seem to work. In so far as the persecutions resemble each other, perhaps it is because of the nature of persecution itself. Is persecution in science different from religious or ideological persecution at all?

Most of the examples that we have given involve rather few people. In the case of Copernicus, for instance, only four people's books were specifically condemned: Copernicus himself (posthumously), Diego de Zuniga, the Carmelite Paolo Foscarini, all in 1616, and Galileo in 1633. Religious persecution, however, has involved hundreds or thousands of people at a time. Sixty thousand Huguenots suffered forcible conversion in three days in 1684 [3]. Similarly with racial persecution: seven million Jews perished in the Holocaust.

This difference may perhaps be explained by the very limited degree of commitment to science among the general public. Few people have gone to the stake for it, or been devoured by lions for it. This is in contrast with the widespread effect, for instance, of a particular interpretation of the Bible in the sixteenth century or widespread hatred against a particular race in the twentieth. A feature of persecution in science is its organized character compared with the uncontrolled fury of mass religious persecution or of 'ethnic cleansing'.

Although scientific persecution may affect a limited number of people directly, however, its indirect effect may last for hundreds of years, as in Italian science after Galileo. Though as far as fashionable Italian society was concerned, the Galileo case was only of passing interest. In all these kinds of persecution, however, the authorities, secular or ecclesiastical, tend to be the prime movers, though they may be stung into activity by the victim.

Another reason for the relatively small scale of persecution in science is that the temporal ruler's interests are seldom involved to any great extent in the controversy. Where security arises, as in the Rosenberg case, intervention is more draconian. This may have been expected in the Oppenheimer case also, but here, as we have seen, he was non-torturable, and was not viewed by anyone as a real threat. His trial was perhaps largely a deterrent to others, in so far as it can be explained at all. More probably, it may simply have arisen from the lack of proportion and irrationality that seems to beset some security services. In most establishments that do some classified work, there is friction between the scientists and the security staff, amounting sometimes to antagonism. Both sides doubt the competence of the other. The scientists doubt the intelligence, or even the sanity, of the security staff. The view of the security staff, usually military or ex-military, was well put by Colonel Lansdale at the Oppenheimer enquiry [4]:

> The scientists *en masse* presented an extremely difficult problem. The reason for it, as near as I can judge, is that with certain outstanding exceptions they lacked what I called breadth. They were extremely competent in their field but their extreme competence in their chosen field led them falsely to believe that they were as competent in any other field.

The suggestion that soldiers are broader in their outlook than scientists would be treated as lunacy by scientists. Certainly neither Galileo nor Oppenheimer could be described as narrow in outlook, in any normally accepted meaning of the term.

We have not yet considered in this section the material of Chapter 5. Though there were undoubtedly persecutions, the most crucial issue, the controversy between Boltzmann and the German chemists, appears at first sight to have been a dialogue or at most an equal conflict. That it was in fact a persecution is difficult to prove. Obviously, it was a very different kind of experience from those of Galileo and Oppenheimer. They, however, were supported by the overwhelming majority of their friends, whereas Boltzmann was isolated, apart from junior assistants. His former disciples had turned away from him, if not personally, at least from his physics, and Boltzmann was extremely torturable.

13.6. Final conclusions

From all that has been said in this book, we conclude the following.

1. In the cases we have considered, where an eminent individual is persecuted by an institution, the sequence of events is similar, whether the context is religious or ideological. This is particularly true for Galileo and Oppenheimer, and more obscurely, for Roger Bacon. The exception to this is Bruno, but he did not have international standing.
2. We find no evidence for malign conspiracies in the cases we have considered, though hostility is present and may be strong. In cases where an individual is in conflict with his colleagues, e.g. Boltzmann and Einstein, the pressures may amount to persecution; though they are not recognized as such by the majority group, but are rather seen simply as a search for truth.
3. An individual is protected from savage retribution if he has:
   (a) a powerful protector (in Oppenheimer's case, his reputation with the general American public);
   (b) an international reputation;
   (c) advanced age.

   If these conditions are met the sentence will be, by the standards of the time, lenient.
4. In the cases of Boltzmann and Einstein the strength of feeling on both sides seems to have been greater than amongst members of the Inquisition's commission or of the Gray tribunal, which were juridical in character.
5. In the Galileo case, it is perfectly plausible to argue that the persecution was really directed at Riccardi, who had given one of the five imprimaturs for the *Two Worlds Dialogue*.

   Alternatively, it can also be argued that the main object of the persecution was to prevent Galileo from publishing further on atomism. *Il Saggiatore* could hardly be condemned, since it had been praised by Urban VIII. In view of the extreme sensitivity of the theory in eucharistic theology at the time, this possiblity can certainly not be ruled out. Equally, however, it would be extremely difficult to prove, unless further documents came to light. A search for such documents would be well worth while, though if they had still existed it is probable that Favaro would have found them.

POSTSCRIPT

Workers with the Hubble Telescope and with ground-based telescopes, observing supernovae at great distances from us, have recently announced strong evidence that the universe is open and will expand indefinitely (*Science* **279** 651, 30 January 1998). Moreover, there is some evidence for a repulsive force which is enhancing the universal expansion.

If these results are confirmed, we are witnessing the only performance of the universe's life, from the intensive heat and pressure of the big bang to the cold and remote depths of space.

May 1998

REFERENCES

Chapter 1. Introduction

1. Bridges J H 1914 *The Life and Work of Roger Bacon* (London: Williams and Norgate) p 105
2. Kuhn T 1977 *The Essential Tension* (Chicago: Chicago University Press) p 131
3. Medawar P 1967 *The Art of the Soluble* (London: Methuen)
4. Kuhn T *op. cit.* p 149

Chapter 2. Hypatia, Bacon and Bruno

1. Dzielska M 1995 *Hypatia of Alexandria* (Cambridge, MA: Harvard University Press) p 68
2. Dzielska M *op. cit.* p 51
3. Dzielska M *op. cit.* p 25
4. Jeans J 1944 *The Growth of Physical Science* (Cambridge: Cambridge University Press) p 101
5. *Dictionary of National Biography: Bacon R* p 847
6. Bridges says:

   What were the novelties that constituted his crime we do not know. It was not perhaps difficult to show that he had gone too far in connecting changes in religious faith with conjunctions of Jupiter and Mercury.

   Bridges J H 1914 *The Life and Work of Roger Bacon* (London: Williams and Norgate)
7. Bacon's credulity is illustrated by many passages in the *Opus Majus*, e.g. on pp 598–615 he has a serious and scientific description of the spectrum:

   In a similar way also he will be able to test the shape in which the colours were dispersed. For by means of the crystalline stone [Iceland spar] and substances of this kind he will find the shape straight. By means of the eyelids and eyebrows and by many other means and also by means of holes in rags he will observe whole circles coloured.

This sensible discussion of interference and diffraction effects appears on p 589, but on p 623 there is the following:

It is certain that wise men of Aethiopia have come to Italy, Spain, France, England and those lands of the Christians in which there are good flying dragons and by the secret arts they possess they lure the dragons from their caverns. They have saddles and bridles in readiness and they ride on these dragons at high speed, so that the rigidity of their flesh may be overcome and its hardness tempered, just as in the case of boars and bulls that are driven about by dogs and beaten in various ways before they are killed for food. After they have domesticated them in this way they have the art of preparing their flesh, similar to the art of preparing the flesh of the Tyrian snake, and this prolongs life and sharpens the intellect beyond all conception.

8. *Encyclopaedia Britannica: Bacon R* p 568
9. Lindberg D C 1983 *Roger Bacon's Philosophy of Nature* (Oxford: Clarendon) p lv
10. Bridges J H *op. cit.* p 49
11. Bridges J H *op. cit.* p 151
12. Lindberg D C *op. cit.* p 185
13. Lindberg D C *op. cit.* p 140
14. Bruno G 1964 *The Expulsion of the Triumphant Beast* Engl. Transl. Imerti A (New Brunswick: Rutgers University Press)
15. *Ibid* p 15
16. *Ibid* p 49
17. *Ibid* p 63
18. Bruno G 1584 *Universo e il mondo infinito* (Venice)
19. Bruno G *The Expulsion of the Triumphant Beast* Introduction *op. cit.*
20. Bruno G *op. cit.* p 255
21. Bruno G *op. cit.* p 121
22. Rowan-Robinson M 1977 *Cosmology* (Oxford: Oxford University Press) p 64

Chapter 3. Galileo and the Inquisition

Primary sources in these references are taken from *The Galileo Affair*, a documentary history with letters and shorter documents [Finocchiaro M A 1989 *The Galileo Affair* (Berkeley, CA: California University Press) and translated from the National Edition by him].

1. Favaro A 1914 *Two New Sciences Dialogue* Engl. Transl. Crew H and de Salvio A (New York: Dover) pp ix–x

2. Wallace W 1984 *Galileo and his Sources* (Princeton, NJ: Princeton University Press)
3. Russell H, Dugan R and Stewart J 1945 *Astronomy* vol 1 (Boston: Ginn) p 316
4. Caccini's deposition to the Inquisition, 20 March 1615:

   There appeared personally and of his own accord at Rome in the great hall of examinations in the palace of the Holy Office, in the presence of the Reverend Father Michelangelo Seghizi OP ... Father Tommaso Caccini ... I had spoken with the most illustrious Lord Cardinal Aracoeli about some things taking place in Florence, and yesterday he sent for me and told me that I should come here and tell you everything ... on the fourth sunday of Advent of this past year I was preaching at the church of Santa Maria Novella in Florence ... and I began with the story of Joshua begun earlier. Precisely on this Sunday I happened to read the passage of the tenth chapter of that book where the sacred writer relates the great miracle which God made in answer to Joshua's prayers by stopping the sun, namely 'Sun stand thou still on Gibeon' etc. After interpreting this passage first in a literal sense and then in accordance with its spiritual intention for the salvation of souls, I took the opportunity to criticize, with that modesty which befits the office I held, a certain view once proposed by Nicolaus Copernicus and nowadays held and taught by Mr Galileo Galilei, mathematician, according to public opinion very widespread in the city of Florence. This is the view that the sun is immovable ... Although this charitable warning greatly pleased many educated and devout gentlemen, it displeased certain disciples of the above mentioned Galileo beyond measure; thus some of them approached the preacher at the cathedral so that he would preach on this topic against the doctrine which I had expounded...

   Finocchiaro M *op. cit.* p 136
5. Redondi P 1987 *Galileo Heretic* (Princeton, NJ: Princeton University Press) p 150
6. Ogg D 1948 *Europe in the Seventeenth Century* (London: Black) p 369
7. de Santillana G 1958 *The Crime of Galileo* (London: Heinemann) p 326
8. Letter to the Grand Duchess Christina:

   As your most Serene Highness knows very well, a few years ago I discovered in the heavens many particulars which had been invisible until our time. Because of their novelty, and because of some consequences deriving from them which contradict certain physical propositions commonly accepted in philosophical schools, they roused against me no small number of professors, as if I had placed these things in heaven with my hands in order to confound nature and the sciences ... they produced various matters, and they published various writings full of useless discussions and sprinkled with quotations from the Holy Scripture, taken from passages which they do not properly understand and which they inappropriately adduce. This was a very serious error and they might not have fallen into it had they paid attention to St Augustine's very useful advice concerning how to proceed with care in reaching definite decisions about matters which are obscure and difficult to understand by means of reason alone. For, speaking about a particular physical conclusion pertaining to heavenly bodies, he writes this: (On the literal interpretation of Genesis, book 2, at the end): 'Now then always practising a pious and serious moderation, we ought not to believe anything lightly about an obscure subject, lest we reject, out of love for our error, something which later may be truly shown not to be in any way contrary to the holy books of either the Old or the New Testament'.

9. de Santillana G *op. cit.* p 99

10. Brodrick J 1964 *Galileo* (London: Chapman) pp 105–6

11. Brodrick J *op. cit.* pp 106–7

12. Brodrick J *op. cit.* p 108

13. Brodrick J *op. cit.* p 115

14. Anonymous complaint about *The Assayer* submitted to the Inquisition in 1624 or 1625, p 202:

Having gone through in the past few days the book by Mr Galileo Galilei entitled *The Assayer*, I came to consider a doctrine once taught by some ancient philosophers, but effectively refuted by Aristotle, though refurbished by the same Mr Galileo [summarizes the atomist point of view] ... according to Anaxagoras and Democritus, a consecrated host will have the substance of bread no less than an unconsecrated host given that according to them a corporeal substance consists of an aggregate of atoms arranged in this or that manner...

15. de Santillana G *op. cit.* p 183

16. de Santillana G *op. cit.* p 184

17. de Santillana G *op. cit.* p 185

18. Niccolo Riccardi, Vatican Secretary, to the Florentine Inquisitor, 24 May 1631, p 212:

... due to current restrictions on the roads and the risks for the manuscript, and since the author wishes to complete the business there, Your Very Reverend Paternity can avail of your authority and dispatch or not dispatch the book without depending in any way on my review.

19. de Santillana G *op. cit.* p 185

20. Niccolini to Galileo, 19 July 1631, p 214:

My Very Illustrious and Most Honourable Sir: After an infinity of cares, finally the preface to your distinguished work has been corrected, as you can see from the enclosed note addressed to the

Father Inquisitor ... I rejoice with you for the termination of this business.

Riccardi to the Florentine Inquisitor, 19 July 1631, p 213:

In accordance with the order of Our Master about Mr Galilei's book, besides what I mentioned to Your Reverend Paternity regarding the body of the work, I send you this beginning or preface to be placed on the first page; the author is free to change or embellish its verbal expressions as long as he keeps the substance of the content. The ending should be on the same theme.

21. Niccolini to the Tuscan Secretary of State, 11 September 1632, p 232:

I have shared with the Master of the Sacred Palace the contents of your letter dated the 30th of last month, in regard to the business of Mr Galilei. I decided to do this not so much because of the friendliness and trust which exists between us, but more because of what the Pope told me at the last audience on this matter, as I also reported in previous letters. He answered and advised that I should express the feelings of His Most Serene Highness if we want to ruin Mr Galilei and break with His Holiness, but that if we want to help him we should completely abandon such expressions of complaints...

Thus Florentine impatience should not be voiced, but equally the Pope could not afford to alienate Florence, an ally already veering towards Spain; so caution on both sides was necessary.

22. Firenzuola to Cardinal Barberini, 28 April 1633, p 276:

I proposed a plan, namely that the Holy Congregation grant me the authority to deal extrajudicially with Galileo, in order to make him understand his error and once having recognised it to bring him to confess it ... In order not to lose time, yesterday afternoon I had a discussion with Galileo and after exchanging innumerable arguments and answers, by the Grace of the Lord I accomplished my purpose: I made him grasp his error so that he clearly recognised that he had erred and gone too far in his book...

23. The ten cardinals pronounced sentence of perpetual house arrest. They were: Gasparo Borgia, Felice Centini, Guido Bentivoglio, Desiderio Scaglia, Antonio Barberini, Zacchia, Gessi, Verospi, Francesco Barberini and Marzio Ginetti. Borgia, Francesco Barberini and Zacchia refused to sign the condemnation of 22 June 1633.

24. In his abjuration, which was pathetic, Galileo agrees that he had been told to completely abandon the Copernican hypothesis, although a few weeks earlier he had denied this. We can assume that the abjuration was at least in part written for him.

25. de Santillana G *op. cit.* preface, pp vii–viii

26. de Santillana G *op. cit.* p 129

27. Brodrick J *op. cit.* p 109

Chapter 4. Kepler

1. The most potent single factor in his early enthusiasm for Copernicanism appears to be found in its exaltation in the dignity and importance of the sun. Founder of exact modern science that he was, Kepler combined with his exact methods and indeed found his motivation for them in certain long discredited superstitions, including what is not unfair to describe as sun worship.

   Burtt E A 1954 *The Metaphysical Foundations of Modern Science* (New York: Doubleday Anchor Books) p 58
2. Newton, in a characteristically caustic remark, said:

   Kepler knew ye orb to be not circular and guest it to be elliptical.

   Turnbull H W (ed) 1960 *The Correspondence of Isaac Newton* (Cambridge: Cambridge University Press) vol 2, p 436.
3. J Kozamthadam points out that the discovery took years of calculation. Kozamthadam J 1994 *The Discovery of Kepler's Laws* (Notre Dame) p 215
4. Boas M 1962 *The Scientific Renaissance 1450–1630* (London: Collins) p 287
5. Kepler introduces the idea of a *species* emanating from the sun and causing the motion of the planets. The idea is reminiscent of Roger Bacon's, but so far as we know Kepler was not acquainted with Bacon's work.

   Once I believed that the cause which moves the planets was precisely a soul, as I was of course imbued with the doctrine of J C Saliger on moving intelligences. But when I pondered that this moving cause grows weaker with distance, and that the sun's light grows thinner with distance from the sun, from that I concluded that this force is something corporeal, that is, an emanation which a body emits, but an immaterial one

   Kepler J 1981 *Mysterium Cosmographicum* Engl. Transl. Duncan A M (New York: Abaris Books) p 203
6. Boas M *op. cit.* p 287
7. Knight D C 1965 *Johannes Kepler* (London: Chatto and Windus)
8. Ogg D 1948 *Europe in the Seventeenth Century* (London: Black) p 88
9. Ogg D *op. cit.* p 144

Chapter 5. Atomic Theory and Positivist Philosophy

1. Brush S 1983 *Statistical Physics and the Atomic Theory of Matter* (Princeton, NJ: Princeton University Press) pp 266–7
2. Flowers B and Mendoza E 1970 *Properties of Matter* (London: Wiley) p 165

3. Lord Kelvin 1904 *Baltimore Lectures* (London: Clay) p 184. In these lectures, which were actually delivered in 1884, Kelvin gave credit to Loschmidt and Johnstone Stoney for the idea of measuring the dimensions of molecules:

   A very important and interesting method of estimating the size of atoms, founded on the kinetic theory of gases, was first, so far as I know, thought of by Loschmidt in Austria and Johnstone Stoney in Ireland. Subsequently the same idea occurred to myself and was described in Nature, March 1870 in an article on the size of atoms...
4. Brush S *op. cit.* p 76
5. Brush S *op. cit.* p 76
6. Clark P 1978 Atomism and thermodynamics *Appraisal in the Physical Sciences* Howson M (ed) (Cambridge: Cambridge University Press) pp 41–106
7. Clark P *op. cit.* p 43
8. Clark P *op. cit.* p 84
9. Clark P *op. cit.* p 91
10. Flowers B and Mendoza E *op. cit.* pp 125–6
11. Clark P *op. cit.* p 91
12. Clark P *op. cit.* p 93
13. Weinberg S 1993 *Dreams of a Final Theory* (London: Vintage) pp 140–1
14. Flamm D 1987 Boltzmann's influence on Schrödinger *Schrödinger* C W Kilmister (ed) (Cambridge: Cambridge University Press) pp 4–15
15. Weinberg S *op. cit.* pp 141–2
16. Lord Kelvin *op. cit.* pp 499–502

Chapter 6. N-rays

1. The origin of this story, which has been ascribed to Rayleigh, Kelvin and Michelson, is obscure.
2. 1903 *Nature* **68** 578
3. 1903 *Nature* **69** 47
4. 1903 *Nature* **69** 72
5. 1903 *Nature* **69** 82
6. 1903 *Nature* **69** 119
7. 1903 *Nature* **69** 82
8. 1904 *Nature* **69** 454
9. 1904 *Nature* **69** 454
10. 1904 *Nature* **69** 454
11. 1904 *Acad. of Sci.* 15 February
12. 1904 *Nature* **69** 286
13. 1904 *Nature* **69** 503
14. 1904 *Nature* **69** 335
15. 1904 *Nature* **69** 599

16. 1904 *Nature* **69** 600
17. 1904 *Nature* **70**; *Acad. of Sci.* 16 May
18. 1904 *Nature* **70** 530

Chapter 7. Einstein and the Copenhagen School

1. Many years later, Frank in the article *Einstein, Mach and Logical Positivism* remarked 'roughly speaking we may distinguish according to Max Planck two conflicting concepts in the philosophy of science: the metaphysical and the positivistic. Each of these regards Einstein as its chief advocate and most distinguished witness.'

Schilpp P (ed) 1949 *Albert Einstein Philosopher-Scientist* (New York: Tudor) p 271
2. Pais A 1982 *Subtle is the Lord. The Science and the Life of Albert Einstein* (Oxford: Oxford University Press) p 382
3. Pais A *op. cit.* p 439
4. Pais A *op. cit.* p 445
5. Schilpp P *op. cit.* p 212
6. Einstein A, Podolsky B and Rosen N 1935 *Phys. Rev.* **47** 777
7. Bell J 1987 *Schrödinger* C W Kilmister (ed) (Cambridge: Cambridge University Press) p 46
8. Bell J *op. cit.* p 45
9. Polkinghorne J C 1984 *The Quantum World* (New York: Longmans)
10. Einstein A, Podolsky B and Rosen N *op. cit.* p 779
11. Polkinghorne J C *op. cit.* p 71
12. Sommerfeld A in Schilpp P *op. cit.* p 105:

In spite of all this, in the old question continuum versus discontinuity, he has taken his position most decisively on the side of the continuum. Everything of the nature of quanta, which in the final analysis the material atoms and the elementary particles belong also, he would like to derive from a continuum physics by means of methods which relate to his general theory of relativity...
13. Born M in Schilpp P *op. cit.* p 166

I think that these investigations of Einstein's [1905–6] have done more than any other work to convince physicists of the reality of atoms and molecules, of the kinetic theory of heat, and of the fundamental part of probability in the natural laws. Reading these papers one is inclined to believe that at that time the statistical aspect of physics was preponderant in Einstein's mind; yet at the same time he worked on relativity where rigorous causality reigns. His conviction seems always to have been, and still is today, that the ultimate laws of nature are causal and deterministic, that probability is used to cover our ignorance if we have to do with numerous particles, and that only the vastness of this ignorance pushes statistics to the forefront.

14. Margenau in Schilpp P *op. cit.* p 251

In a number of places Einstein expresses his indebtedness to Mach and it is easy to trace his concern over observability to the philosopher.

15. Bohr N in Schilpp P *op. cit.* p 208

It is decisive to recognise that, however far the phenomena transcend the scope of classical physical explanation, the account of all evidence must be expressed in classical physical terms.

16. Aspect A, Dalibard F, Grangier P and Roger A 1981 *Phys. Rev.* **47** 460

17. Pais A *op. cit.* p 185

Chapter 8. Oppenheimer and the AEC

1. Rhodes R 1986 *The Making of the Atomic Bomb* (London: Penguin) p 446
2. Groves L 1963 *Now it Can be Told* (London: Deutsch) p 60
3. Groves L *op. cit.* p 111
4. Rhodes R *op. cit.* p 467
5. Rhodes R *op. cit.* p 536
6. Rhodes R *op. cit.* p 570
7. Rhodes R *op. cit.* p 769
8. Bethe H 1950 *Sci. Am.* April
9. Reid R W 1971 *Tongues of Conscience* (London: Panther) p 282–4
10. 1954 *The Trial: In the Matter of J Robert Oppenheimer. Transcript of Hearings before the Personnel Security Board, US Atomic Energy Commission* (Washington, DC: USAEC) p 710 (referred to as Transcript)
11. Transcript *op. cit.* p 726
12. Reid R W *op. cit.* p 289
13. Dyson F 1993 *From Eros to Gaia* (London: Penguin)
14. Transcript *op. cit.* p 262
15. Reid R W *op. cit.* pp 297–8
16. Transcript *op. cit.* p 21
17. Transcript *op. cit.* p 22
18. Dyson F *op. cit.* p 255

Chapter 9. Blackett, Cherwell, Tizard

1. Waugh E 1987 *Men at Arms* (London: Penguin) p 12
2. Giovannitti L and Freed E 1965 *The Decision to Drop the Bomb* (London: Methuen)

3. Gill R 1985 *A Textbook of Christian Ethics* (Edinburgh: Clark) p 304ff
4. Aquinas Thomas *Summa Theologia* 2–2ae xl.1 (Engl. Transl. Gilby T 1951 *Philosophical Texts* (Oxford: Oxford University Press) p 348)
5. Giovannitti L and Freed E *op. cit.*
6. Snow C P 1971 *Science and Government in Public Affairs* (London: Macmillan) p 99
7. Snow C P *op. cit.* p 125–6
8. Blackett P M S, quoted by Snow C P *op. cit.* p 179
9. Irving D 1963 *The Destruction of Dresden* (London: Kimber) p 237
10. Frankland N 1965 *The Bombing Offensive Against Germany* (London: Faber and Faber) pp 72–3
11. Liddell Hart B 1970 *History of the Second World War* (London: Cassell) pp 596–7
12. Irving D *op. cit.* p 36
13. Liddell Hart B *op. cit.* p 141
14. Churchill W S 1951 *History of the Second World War* vol 3 (London: Cassell) p 451
15. Liddell Hart B *op. cit.* p 599
16. Irving D *op. cit.* p 73
17. Walzer M 1977 *Just and Unjust Wars* (New York: Basic Books) pp 153–5
18. Anscombe E, quoted in Rhodes R 1986 *The Making of the Atomic Bomb* (London: Penguin) p 698
19. Walzer M *op. cit.* p xvi
20. Gill R *op. cit.*
21. Bainton R H 1961 *Christian Attitudes to Peace and War* (London: Hodder and Stoughton)
22. Flannery A (ed) 1975 *Gaudium et Spes* (Dublin: Dominican) no 80
23. Bonhoeffer D 1966 *Ethics* (London: Fontana) p 159
24. Feis H 1961 *Japan Subdued* (Princeton, NJ: Princeton University Press)
25. Dyson F 1992 *From Eros to Gaia* (London: Penguin) p 258–9
26. Churchill W S *op. cit.* vols 4, 5
27. Doenitz K 1943, quoted in Arnold-Foster 1973 *The World at War* (London: Collins) p 80

Chapter 10. Big Bang versus Continuous Creation

1. Burbidge G 1990 *Physical Cosmology* (Paris: Editions Frontieres) p 58
2. Burbidge G *op. cit.* p 74
3. Sciama D W 1971 *Modern Cosmology* (Cambridge: Cambridge University Press) p 116
4. Silk J 1989 *The Big Bang* (San Francisco: Freeman) p 61
5. Sciama D W *op. cit.* p 85
6. Contopoulos G and Kotsakis D 1984 *Cosmology* (Berlin: Springer) p 139

7. Arp H C 1987 *Quasars, Redshifts and Controversies* (Berkeley, CA: University of California Press)
8. Burbidge G *op. cit.* p 67

Chapter 11. Multiple versus Plural

1. Heitler W and Janossy L 1949 On the absorption of meson-producing nucleons *Proc. Phys. Soc.* **72** 374
2. Heisenberg W 1955 *Nuovo Cimento Suppl.* **2** 97 (A description of the multiple theories, but particularly that of Landau. Some information is also given on experimental results.)

Chapter 12. Missing Magnetic Monopoles

1. Dirac P A M 1931 *Proc. R. Soc.* **133** 60
2. Dirac P A M 1948 *Phys. Rev.* **74** 817
3. Malkus W V R 1951 *Phys. Rev.* **83** 899
4. Porter N A 1960 *Nuovo Cimento* **16** 958
5. Polykov A, t'Hooft G and Rubakov V 1981 *JEPT Lett* **33** 644
6. Calvera B 1982 *Phys. Rev. Lett.* **88** 1378

Chapter 13. Conclusions

1. de Santillana G 1958 *The Crime of Galileo* (London: Heinemann) preface, pp viii–ix
2. de Santillana G *op. cit.* p 288
3. Christie-Murray D 1976 *A History of Heresy* (Oxford: Oxford University Press) p 178
4. Reid R W 1971 *Tongues of Conscience* (London: Panther) p 295

GLOSSARY

AEC Atomic Energy Commission.

Alamogordo Site of the first nuclear weapon test in 1945.

Albigensians Sect concentrated in the south of France in the thirteenth century. Held that there were principles of good and evil, and other heresies. Persecuted by the Church.

Alexander Macedonian king and general. His conquests extended as far as India. Died in 323 BC.

Alexander of Hales Philosopher and theologian who taught in the University of Paris in the mid-thirteenth century. Roger Bacon attended his lectures.

Alexandria City founded by Alexander in the Nile delta just before his death. Its university, the Museum, with an enormous library, was active until it was suppressed by the Arabs in 640 AD.

Alhazen Arab scientist and mathematician of the tenth century. Worked on optics, particularly on reflection and refraction from curved surfaces.

Allison, Samuel American physicist. Working on the uranium project from early in the Second World War.

Almagest The great textbook of astronomy constructed by Claudius Ptolemy, c150 AD. The most influential astronomy text for over a thousand years, until it was overthrown by Copernicus in the sixteenth century. Held that the stars, Sun and planets rotate about an immovable Earth.

Alsace Province of France held at times by the Germans. Its ownership was a contributory cause of several wars.

Amaury or Almaric of Bena Pantheist philosopher who held that God, matter and mind had the same substance. Condemned by the Pope, his body was disinterred and buried in unconsecrated ground in 1209 AD.

America The source of much wealth in the sixteenth and seventeenth centuries, particularly in Spain. This tended to destabilize the economies of European countries.

Anaxagoras (c500–428 BC) Athenian philosopher. Held that the Sun, Moon and stars were heated objects. Condemned for this and for other doctrines. Held that matter was constructed of infinitesimal particles, created by mind.

Anaximander (c611–546BC) Greek philosopher born in Miletus. Succeeded Thales and developed some of his ideas. Held that the Earth is at the centre of the universe, unsupported.

Andrews, Thomas (1813–85) Irish physicist. Well known for his work on liquefaction of gases, and on deviations from the ideal gas laws.

Andromeda Nebula A large galaxy close to our own, but containing more stars.

Annus Mirabilis The name often given to 1905, when Einstein produced his three epoch-making papers; on the photoelectric effect, Brownian motion, and special relativity.

Anscombe, Elisabeth A professor of moral philosophy at Oxford, who produced a famous pamphlet on total war, when it was proposed that Mr Truman should be given an honorary degree.

Anti-semitism A racial prejudice particularly widespread in central and eastern Europe in the first half of the twentieth century.

Apollonius (250–220 BC) Greek mathematician. Lived in Alexandria. Famous for some theorems in geometry, but historically most important for his work on conic sections.

Aquinas, Thomas (1225-74 AD) The greatest of the scholastic philosophers. His teaching, adapted from Aristotle, dominated the Roman Church until almost the present day.

Arab Empire A civilization which lasted a thousand years, covering an area from India to Spain. The Arabs were expelled from Spain in the sixteenth century. Latin translations of many Greek authors entered Europe through this route.

Archimedes (287–212 BC) The greatest mathematician and physicist of antiquity. Made contributions in many fields: geometry, hydrostatics, levers, burning mirrors.

Aristarchus (c270 BC) One of the greatest Greek astronomers. Measured the distance from the Earth to the Sun for the first time. Believed that the Earth and the planets rotated about the Sun. His work was not accepted.

Aristotle (384–322 BC) The most influential Greek philosopher in both the Arab and the western world. Initially his work caused progress in science, particularly biology, but by the seventeenth century had become sterile and tedious. It was swept from science but remained influential in theology.

Army Intelligence The branch of the United States Army attached to the Manhattan Project for security purposes. Notorious for its investigation of J R Oppenheimer.

Arp, H American astronomer. Over many years he found associations of galaxies with differing red-shifts. His results are controversial, but if true are of great importance.

Astrology The view that the planets and the stars can affect our lives. Belief in this view lasted for thousands of years, and while apparently without foundation, provided a living for many astronomers. Thus it encouraged rather than retarding real progress.

Atomic bombs Weapons depending on a chain reaction in uranium 235 or plutonium 239. Early versions released the equivalent of 20 000 tons of TNT.

Atomism A model originating with the Greeks, Leucippus and Democritus in the fifth century BC. Matter was said to be composed of small indestructible particles. This view triumphed in the nineteenth century, only to be followed by the disintegration of the atomic nucleus in the twentieth.

Augsburg, Peace of An agreement reached in 1555, whereby the religion of the ruler of a state determined the religion of the people.

Augustine With Thomas Aquinas, one of the two great doctors of the Western Church. His views on the origin of time are still of scientific interest.

Averroes Latinization of Ibn Rushd. (1126–1198). An important but controversial commentator on Aristotle. His views were influential up to the seventeenth century. They included an eternal universe, and opinions on religion suited to the common people. His teachings influenced Roger Bacon, Giordano Bruno and even Galileo, but were repeatedly condemned by the Church from 1216 onwards.

Avogadro, Amadeo (1776–1856) Italian physicist born at Turin. Formulated the concept of a mole of different substances each containing the same number of molecules.

B29 American bomber aircraft capable of carrying twenty tons of high explosives. Nicknamed the 'Superfortress', it had heavy armament and long range. It was brought into use by the US Air Force in the last year of the Second World War, mainly in attacks on the Japanese mainland. It was used in the incendiary raids on Tokyo in March 1945, and also carried the Hiroshima and Nagasaki bombs.

Babylonian mathematics They had a relatively simple positional system with a modulus of 60. This enabled them to do calculations impossible in the ancient world.

Bacon, Francis English philosopher. His views on induction have been very influential in the philosophy of science, but have rather lost ground recently.

Bacon, Roger One of the subjects of our extended treatments. His work on optics alone establishes him as a major figure, but his life remains somewhat of an enigma.

Baldwin, Stanley (1867–1947) British politician and Prime Minister. His belief that a bomber would overcome all defences was influential in forming policy, but fortunately not conclusively so.

Balmer series A regular sequence in the spectrum of hydrogen, discovered by a Swiss schoolmaster in the 1880s. Played an important part in Bohr's theory of the hydrogen atom.

Barberini, Maffeo (1568–1644) Seventeenth century Italian cardinal who became Pope Urban VIII. Having been a friend of Galileo he presided over his condemnation in 1633.

Bard, Ralph Spokesman for the US Navy on the interim committee set up in 1945. He was the only person to vote against the decision to use the atomic bomb on a mixed military and civilian target.

Bargaining The process of abandoning part of a theory in return for the acceptance of another part. While it is common in other kinds of conflict, we argue that it is rare in science.

Barycentre The centre of gravity of the solar system. It lies within the body of the Sun.

Bell, Bishop George Anglican Bishop of Chichester during the Second World War. Questioned the wisdom and morality of Allied bombing policy. A friend of Bonhoeffer and other Germans opposed to Hitler.

Bell, John British physicist born in Belfast. Explored the implications of the Einstein, Podolsky and Rosen quantum paradox.

Bellarmine, Cardinal Robert (1542–1621) Prominent Jesuit member of the Vatican Curia when Galileo visited Rome in 1616.

Berkeley, California University town. Main physics activities in the 1940s centred around large cyclotrons.

Bernoulli, Daniel (1700–82) One of a distinguished family of Swiss scientists. Explained the anomalies in the kinetic properties of gases in terms of a a finite diameter for the molecules.

Bessel functions Mathematical functions particularly well suited for problems with cylindrical symmetry. Their inventor also worked in astronomy and was famous for his measurement of stellar paradox in 1838.

Bethe, Hans German theoretical physicist. Did important work on electron collisions in the 1930s. Emigrated to the USA and became Head of the Theoretical Physics Division at Los Alamos.

Bhabha, H Distinguished Indian physicist. Worked on the theory of electron–photon cascades in the 1930s. Later became a senior figure in physics in India.

Birkbeck College, University of London Traditionally has specialized in courses for those who otherwise could not go to university.

Bisley, Gloucestershire Tradionally the birthplace of Roger Bacon.

Blackett, P M S (1897–1974) British physicist. Began his career in the Navy, afterwards going to Cambridge, Birkbeck College, Manchester and Imperial College. Specialized in cloud chambers, operational research and palaeomagnetism. Nobel Prizewinner 1948.

Blitzkrieg Very rapid movement of armoured columns executing flanking movements. Used by the Germans in 1940.

Blondlot, Rene French physicist. Believed that he had discovered a new form of radiation called N-rays, which were afterwards found to be spurious.

Bluffing Common in military and other forms of conflict. Does not appear to be significant in scientific conflict.

Bohemia Part of the Czech Republic. A centre of trouble and rebellion during the Thirty Years War.

Bohr, Niels (1885–1962) Danish physicist. Developed the first viable theory of the hydrogen atom. A prominent figure in world physics; engaged in a long-standing controversy with Einstein.

Boltzmann, Ludwig (1844–1906) Austrian physicist. Pioneer of kinetic theory and statistical mechanics.

Bomber Command Division of the RAF concerned with bombing during the Second World War.

Bonaventure (1221–74) Franciscan theologian. Became general of the order during Roger Bacon's imprisonment.

Bondi, Sir Hermann Cosmologist and strategist of science. With Hoyle and Gold, he developed the steady-state theory of the universe.

Bonhoeffer, Dietrich German moral philosopher and theologian. Opposed Hitler's policies and was executed by the Nazis in 1945.

Borgia, Gaspare One of three cardinals who voted against the condemnation of Galileo.

Born, Max A leading figure in the development of the probabilist interpretation of quantum mechanics.

Bose Indian theoretical physicist. Made important advances in quantum statistics.

Boyle's Law A proportionality between the pressure of a gas and the reciprocal of its volume. Discovered by Boyle and Towneley in the seventeenth century.

Brahe, Tycho Sixteenth century Danish astronomer. Opposed the Copernican model because it did not agree with his observations. Developed his own model of the system.

Brandenburg State in eastern Germany which became part of Prussia.

Breit, G American physicist working on the Manhattan Project in Chicago and Los Alamos.

Bremsstrahlung Emission of a high-energy photon when an electron is decelerated. Important in cascade showers.

Bristol University Scene of important discoveries in particle physics in the late 1940s and early 1950s, under the leadership of C F Powell.

de Broglie, Prince Louis (1892–1987) French physicist. Discovered wave–particle duality.

Brownian motion Random motion of small particles in a liquid. In 1905 subject of a theory by Einstein which demonstrated the existence of atoms.

Bruno, Giordano (1548–1600) Italian philosopher executed for heresy in Rome. First person to express the Copernican principle.

Bryan, William Jennings (1860–1925) American populist leader. Nominated three times unsuccessfully for the Presidency by the Democrats.

Bush, Vannevar (1890–1974) Inventor of the analogue computer and senior university administrator. Served as a member of several important committees in the Manhattan Project.

Butow, Robert American historian who spent several years in Japan, investigating the closing stages of the Second World War and the Japanese surrender.

Butterfield, Herbert Professor of Modern History at Cambridge. Wrote an important book on the scientific revolution in 1948.

Byrnes, James Assistant to President Truman, he deputized for him in a number of important committees on the use of the atomic bombs.

Caccini, Tommaso Florentine Dominican, violent enemy of Galileo and the Copernican model. Denounced Galileo to the Inquisition in 1615.

Calendar, Julian The Egyptian civil calendar was deficient by one quarter of a day. The Julian calendar, adopted in the first century BC, added 1/4 of a day by putting in a leap year every four years. This was sufficient for 1500 years but was then replaced by the Gregorian calendar, which drops the leap year for centennial years, with some small exceptions.

Campo di Fiori The square where Bruno was burnt to death. It now has a massive statue of him.

Carlson, J F American physicist who worked with Oppenheimer in the 1930s on the theory of cascade showers.

Casablanca Conference In January 1943, Churchill, Roosevelt and their generals attended a conference which decided a number of military and political issues.

Cassiopeia A One of the strongest radio sources in the sky, it is a supernova remnant.

Castelli, Benedetto Professor of Physics at Pisa in the early years of the seventeenth century, he was a close friend of Galileo. He was a Benedictine monk.

Cavity magnetron The device which made it possible to generate megawatt pulses of microwaves. Developed by Randall and Boot at the Physics Department of Birmingham University in 1940.

Celestial matter According to Aristotle, a more perfect and ethereal material than the gross substance found on Earth.

Cena di Ceneri Title of one of Bruno's books set on Ash Wednesday at an English country house. It is a powerful defence of the Copernican model.

Centrifuging A possible method of separating U235 from U238. Rejected by the Americans in the Second World War, it was successfully developed by the Germans after the war for peaceful purposes.

Cepheid variables Stars whose intrinsic brightness is related to the period of variation in brightness. They can be used for distance measurement.

Charge conservation In nuclear reactions or electron–photon interactions, the algebraic sum of the positive and negative charges must remain constant.

Charlemagne (747–814 AD) First Emperor of the Holy Roman Empire.

Cherwell, Lord Title taken by F A Lindemann on his elevation to the peerage.

Cheshire, Group Captain Leonard Founded, organized and led the Pathfinder squadrons in the Second World War bombing raids. Awarded the Victoria Cross.

Chevalier, Haakon A friend of Oppenheimer, he mentioned at a cocktail party that someone he knew was looking for secret information for the Russians. On Oppenheimer's belatedly reporting this to the security authorities, Chevalier lost his job.

Christina of Lorraine, dowager Grand Duchess of Tuscany She was concerned by Galileo's views on Scripture which seemed contrary to the faith. This provoked Galileo's book: *Letter to the Grand Duchess Christina.*

Church, Calvinist The branch of Protestantism which derives from about 1550 and centred on Geneva. Most Presbyterian churches are Calvinist.

Church, Catholic That branch of Christianity which has accepted the authority of the Pope.

Church, Orthodox In communion with the Catholic Church until 1054 AD, the basic theologies of the two Churches are quite similar.

Church, Lutheran The first Reformed Church to successfully break with Rome in 1517, under Martin Luther at Wittenburg where he was Professor of Biblical Studies. In liturgy the new Church had rites similar to Catholic ones. The main theological differences hinged on salvation through Faith alone, and the abolition of indulgences and other less important practices.

Cistercians Order of monks founded by Abbot Robert in 1098. A branch was founded by St Bernard later.

Civilian A category in warfare. Civilians, according to successive Geneva Conventions, may not be attacked unless engaged in war work.

Clausius, Rudolf (1822–88) One of the founders of thermodynamics, proving a number of theorems which led to the second law.

Clavius, Christopher The senior Jesuit astronomer in Rome at the beginning of the seventeenth century. He was open to Copernicus' views, but unfortunately died before they became a matter of controversy.

Clay, J Dutch physicist who demonstrated the charged character of the primary cosmic rays by taking a measuring system to different points on the Earth.

Clayton, William L Assistant US Secretary of State in 1945. Member of the Interim Committee.

Clement IV, Pope Befriended Roger Bacon and asked him to write an encyclopedia of all that was known. He died before the work could be completed.

Clement VIII, Pope (1591–1605) Imposed the full powers of the Inquisition. Bruno was executed during his pontificate.

Collegio Romano University style college run by the Jesuits in the seventeenth century.

Compton, A H American physicist and Nobel Prizewinner. Member of the Scientific Panel in 1945.

Compton, Karl American scientist, administrator and brother of A H Compton. Member of the Interim Committee in 1945.

Conant, James American scientist and president of Harvard. Member of the Interim Committee in 1945.

Conic section Intersection of a cone with a plane in space: circle, ellipse, parabola or hyperbola. Associated with the inverse square law of force, whether gravitational or electrical.

Copernicus, Nicolas (latinized form of Koppernigk) (1473–1543) The accepted founder of modern planetary astronomy. While some Greeks and Arabs had put the Sun at the centre of the solar system, he was the first to do so in mathematical detail.

Copernican Epitome Handbook of Copernican astronomy written by Kepler in three parts 1618–21. Banned by the Inquisition.

Cosmic rays Charged nuclear particles striking the Earth from outer space. Their origin is still unknown.

Cosmic strings Hypothetical nuclear particles, string-like in topology, which may be of fundamental importance in the universe. No experimental evidence has yet been found for their existence.

Council of the Fourth Lateran 1215 Reforming Council which condemned Averroist ideas such as the eternity of the universe, and made a number of political decisions.

Counter-Reformation In the latter part of the sixteenth century, the Roman Church made a number of changes to combat the Reformation. These included the development of the Jesuits and the Council of Trent.

Crab Nebula Remains of a supernova which exploded in 1054 AD. It has been observed at all frequencies from radio up to ultra-high-energy gamma rays. The nebula derives part of its energy from a central pulsar.

Crystalline spheres In some Ptolemaic models the planets are carried on transparent spheres. Some mediaeval writers regarded these as fanciful.

Cube One of the regular solids used by Kepler.

Cuius regio 'Whatever is the religion of the King, will be the religion of the people.' The Augsburg agreement.

Cygnus A Strong radio source whose optical counterpart reveals a peculiar galaxy.

Cyril Patriarch of Alexandria at the time of Hypatia's murder in 415 AD.

Dalén N G Swedish engineer. Awarded the Nobel Prize in Physics in 1912.

Dalton, John (1766–1844) English chemist. Pioneered the scientific atomic theory.

Dartmouth Naval College Training institution for Royal Navy cadets.

Darwin, Charles Founder of the theory of evolution.

David of Dinant Thirteenth century pantheist writer, who believed in the unity of God, matter, soul and mind.

Deceleration parameter, q Measure of the rate at which the universe's expansion is declining. It determines whether the universe is open or closed.

Deferent Main circle of the orbit in the Ptolemaic system.

Degrees of freedom In the context of the kinetic theory of gases, indicates in how many ways a molecule can possess energy.

De Immenso One of Giordano Bruno's tracts against Aristotle.

De Minimo Bruno's tract on atomism.

Democritus (c460-c370) Co-founder of the atomic theory of matter.

De Monade Bruno's work on the monad or atom of everything.

Descartes, Rene (1596–1650) French philosopher and mathematician. Attempted a unified theory of knowledge. Pictured atoms as vortices in a fluid. Invented co-ordinate geometry.

Dialogue on the Two World Systems, or on the chief world systems Galileo's Copernican work of 1631, for which he was condemned.

Diffusion A method of separating the isotopes U235 and U238. Taken up on a large scale in the Manhattan Project.

Diffusion, thermal A possible separation method for U235 and U238. Rejected because of the very large amount of power required.

Dirac, P A M One of the most respected physicists of the twentieth century. Nobel Prizewinner in 1933 for his work on quantum mechanics. Put forward the concepts of the positron and the magnetic monopole.

Dodecahedron Twelve-sided regular solid.

Doenitz, Karl German Admiral commanding the U-boat fleet in the Second World War.

Dominicans Order of friars founded by St Dominic in 1216. For many centuries it was responsible for the Inquisition.

Doppler shift Change in the pitch of a sound, or the wavelength of light, when the source is in motion relative to the observer.

Dowding, Lord Hugh (1882–1970) Commander of Fighter Command in the preparation for, and execution of, the Battle of Britain in 1940.

Dubridge, Lee President of Caltech. Member of the General Advisory Committee of the AEC.

Dukas, Helen Einstein's secretary for many years.

Dulong–Petit Law The law named after its originators in the 1820s which states that the specific heat of a mole tends to three times the gas constant R. Explained by Einstein on the basis of quantum theory.

Duns Scotus (c1265–1308) Scottish Franciscan philosopher. Reacted strongly against Aristotle and the Dominicans.

Earth The third planet of the solar system. In the Ptolemaic model, the centre of the universe.

Earth, age of Can be determined from studies of radioactivity. At one time appeared to be older than the Sun.

Eddington, Sir Arthur (1882–1944) Professor of Astronomy at Cambridge. Worked hard for the acceptance of relativity.

Edison, Thomas Alva (1847–1931) American inventor. Responsible for many patents. Nominated for the Nobel Prize in 1912, but Tesla refused to share it with him.

Egidii, Stefano Florentine Inquisitor at the time of Galileo.

Ehrenfest, Paul (1880–1933) Austrian physicist. Attempted to reconcile the views of Bohr and Einstein, but failed. Committed suicide in 1933.

Einstein, Albert (1879–1955) Probably the greatest physicist of the twentieth century. Made contributions in many parts of physics, but is best known for his theories of special and general relativity. Had a long-running controversy with Bohr over the interpretation of quantum mechanics.

Einstein, Maja Einstein's younger sister.

Electromagnetic separation Process in which the two natural isotopes of uranium are separated in electric and magnetic fields. Taken up on a large scale by the Manhattan Project.

Electron The basic 'atom' of electricity. Negatively charged.

Eltenton, George Shadowy figure associated with the Oppenheimer case. Possibly a Soviet spy.

Elzevir/Elsevier Famous Dutch publishing house, still extant. Published Galileo's *Two New Sciences* dialogue in 1638.

Eniwetok The Pacific island where the first American hydrogen bomb was exploded.

Epicycle In the Ptolemaic model, the small circle attached to the deferent.

Epicureanism A school of philosophy, strongly influenced by atomism. Ethics, physics and logic are the three components of the worldview; ethics being most important.

Equant Point slightly removed from the Earth's centre in the Ptolemaic model. The planet rotates about the equant.

Eratosthenes (c276–194 BC) Alexandrian philosopher and mathematician. Measured the circumference of the Earth with high accuracy.

Ether A theoretical fluid permeating the universe. Easily permeable but highly rigid. It was the medium for carrying light. The need for it was removed by special relativity.

ETH The famous Technische Hochschule in Zurich, where Einstein studied.

Euclid (c300 BC) Greek geometer. Systematized and developed geometry in a form which lasted until the present century.

Euclidean solids Regular solids which Kepler thought could be fitted together to give the sizes of the planetary orbits.

Evans, Ward American chemist, and member of the Gray tribunal.

Evolution The system by which life evolves spontaneously to higher forms. Controversial, particularly in the nineteenth century.

Expanding universe Discovered by Hubble in the 1920s by studies of red-shifts.

Extensive air showers Phenomenon in cosmic rays, in which many particles arrive at the ground simultaneously.

Falklands Scene of the first major naval battle of the First World War.

Farnborough Royal Aircraft Research Establishment.

FBI Federal Bureau of Investigation. The principal Security Agency in the United States.

Ferdinand II Holy Roman Emperor 1619–37.

Ferdinand Grand Duke of Tuscany and patron of Galileo. Reigned 1621–70.

Fermi, Enrico (1901–54) Nobel Prizewinner 1938 for his work on elements produced by slow neutrons. Leader of the team which produced the first self-sustaining chain reaction.

Fighter Command The division of the RAF responsible for defence through fighter aircraft in the Second World War.

Fire storm Combination of intense heat and strong winds, produced by dropping incendiary bombs in high concentration on inflammable targets such as housing.

Firenzuola, Maculano di Commissary of the Inquisition who interrogated Galileo in 1633.

Fission The division of an atomic nucleus into two roughly equal parts.

Fitzgerald, George Francis (1851–1901) Irish physicist.

Florence Capital of the Grand Duchy of Tuscany. Home of Galileo for thirty years.

Florin Gold coin equivalent to the gulden in the seventeenth century. In modern money about fifty dollars.

Flying Fortress The B17. American heavy bomber in the Second World War. Later versions were known as the 'Superfortress'.

Foscarini Antonio Prior of the Carmelites in Naples who wrote a book supporting Copernicus. Condemned by the Inquisition in 1616.

Franciscans Order of mendicant friars founded by St Francis in 1210. Roger Bacon was a member of the order.

Free Silver Early twentieth-century theory that unlimited minting of silver at a fixed ratio to gold would support the economy of the USA.

Friedmann, Aleksander Russian mathematician and relativist.

FRS Fellow of the Royal Society of London.

Fuchs, Klaus (1911–88) Physicist and spy. Worked on the Manhattan Project and at Harwell.

Fusion The combination of deuterium with deuterium or tritium to form helium.

Galaxy Collection, typically of a hundred billion stars, with also magnetic fields, gas and dust clouds.

Galbraith, J K American economist who with Paul Nitze, carried out a survey of Allied bombing policy in the Second World War.

Galileo Galilei (1564–1642) See Chapter 3.

Gassendi (1592–1655) Christian atomist.

Gaudium et Spes (*The Church in the Modern World*) Document of the Second Vatican Council, published 1969. Made uncompromising statements about nuclear weapons.

General Advisory Committee Civilian committee chaired by J R Oppenheimer in the US Atomic Energy Commission.

Gibbon Edward (1737–94) Author of *The Decline and Fall of the Roman Empire*. Sought to demonstrate how classical civilization had been destroyed by Christianity.

Gibbs, Josiah Willard (1839–1903) American physicist. Played a major part in the development of statistical mechanics.

Gilbert, William (1540–1603) Made important advances in the study of magnetism.

Gold, Thomas (1920–). Austrian and British astronomer One of the founders of the steady-state theory of the universe.

Göttingen German university town famous for the development of quantum mechanics.

Grassi, Orazio Italian Jesuit and pamphleteer. Opponent of Galileo.

Graviton The quantum of gravity.

Gray, Gordon Chairman of the personnel security board which judged Oppenheimer's case in 1954.

Graz Town in Austria, home for some years of Kepler.

Gregory IX, Pope (1148–1241) Founder of the Inquisition.

Grienberger Jesuit astronomer in Rome. Friend of Galileo, but sometimes opposed him.

Grosseteste, Robert (c1175–1253) English philosopher and scientist. Chancellor of Oxford University and Bishop of Lincoln.

Groves, General Leslie Director of the Manhattan Project.

Grumman Avenger Fighter-bomber, particularly designed for use on small aircraft carriers. A most successful aircraft.

Guiccardini Florentine ambassador to Rome in 1616.

Gulden Gold coin in circulation generally in Europe in the seventeenth century. See **Florin**.

Gustavus Adolphus (1594–1632) King of Sweden during the Thirty Years War.

Guth, Alan Originator of the theory of universal inflation.

Hanford The site of the large nuclear reactors built for production of plutonium in 1944.

Hapsburgs Family which became virtually hereditary Emperors, although the position was elective.

Harris, Air Chief Marshal Sir Arthur (1892–1984) Commander of Bomber Command of the RAF, 1942–45.

Heisenberg, Werner (1901–76) German physicist. Made some of the most important advances in quantum mechanics.

Heitler, Walter German physicist. Worked on high-energy electron and photon interactions, and later on meson production.

Heliocentric model The model which puts the Sun at the centre of the solar system with the planets rotating about it. Originated with Aristarchus but was first quantified by Copernicus.

Helium, cosmic The fraction of helium formed soon after the big bang, and that made later in stars, are important quantities in cosmology.

Henry III (1207–72) King of England at the time of Roger Bacon. His reign was marked by many wars and disputes.

Henri IV (1553–1610) Originally a Calvinist, but after many vicissitudes became a Catholic in 1593. King of France from 1589. Killed by an assassin.

Herbert, George (1593–1633) English metaphysical and religious poet.

Herschel, John (1792–1871) English astronomer, specializing in extended objects. Worked for a time in South Africa.

Herschel, William (1738–1822) Anglo-German astronomer. Discovered the planet Uranus and some minor members of the solar system.

Hertzsprung, Einar (1873–1967) Danish astronomer. Worked on the stellar distance scale.

Hill, A V (1886–1977) English physiologist, Nobel Prizewinner in 1922. Member of the Tizard Committee.

Hipparchus (c150 BC) Greek astronomer. Produced accurate tables of stars, and made a number of advances in solar astronomy.

Hiroshima Japanese city and military headquarters. Destroyed by a U235 bomb in 1945.

Hiss, Alger American civil servant. Convicted of perjury in spy trial, 1950.

Hobbes, Thomas (1588–1679) English philosopher. Visited Galileo in Florence.

Hoffmann-Stosse Bursts of ionization. Shown by Blackett and Occhialini (1933) to be similar to cascades in a cloud chamber.

Holocaust Murder of millions of Jews by the Nazis during the Second World War.

Holy Roman Empire Founded by Charlemagne, it existed in various forms until 1918, but had long since lost its influence.

Hoover, J Edgar American civil servant. Became dirctor of the FBI in 1924 and remained there until his death in 1972.

Hopkins, Gerard Manley (1884–89) English poet and Jesuit. For seven years he wrote nothing. This silence was broken with *The Wreck of the Deutschland*.

Horrocks, Jeremiah Lancashire astronomer and clergyman. He called attention to the work of Kepler, then little-known. Died in 1641 at the age of 22.

Hoyle, Fred (1915–) One of the originators of the steady-state theory of cosmology, has produced variants on the original. Also carried out important work on nucleosynthesis.

Hubble, Edwin (1889–1953) Discoverer of the cosmological red-shift.

Huguenots French protestants, mainly Calvinist. Left France in 1685 after the revocation of the Edict of Nantes.

Huxley, T H (1825–95) English scientist and campaigner for the theory of evolution.

Huygens, Christiaan (1629–93) Dutch astronomer and physicist. Worked on the wave nature of light and measured the distance to the star Sirius.

Hypatia (c355–415 AD) Greek woman astronomer murdered in 415 AD.

Icosahedron Twenty-sided regular Euclidean solid used by Kepler in his early work on the dimensions of the solar system.

Idealism Philosophical theory that states that the world is not material but an extension of the mind.

Infinite universe First put forward by Giordano Bruno.

Inflation Model of the Universe put forward by Guth in 1981. The universe is thought to have undergone a sudden expansion, very soon after its origin.

Infrared radiation Electromagnetic radiation with wavelength between microwaves and visible light.

Innocent III, Pope (1160–1216) Reforming Pope who called the fourth Lateran council in 1216.

Innocent IV, Pope Pope when Roger Bacon was at Oxford.

Inquisition Organization set up to guard the orthodoxy of the Catholic faith. Founded in 1233 and active in the thirteenth century, it became moribund in the later Middle Ages. Reorganized in Spain as a separate entity and in the rest of Europe in 1543. Active in the Counter-Reformation period.

Interim Committee 1945 Committee set up by the Secretary for War to consider the use of the atomic bomb.

Iraq State set up by the Allies after the First World War, from part of the Ottoman Empire.

Janossy, Lajos Hungarian physicist. Worked in Berlin, Manchester, Dublin and Budapest, specializing in cosmic rays.

Jansky, Karl (1905–50) Founder of radioastronomy.

Jesuits Religious society founded by St Ignatius Loyola in 1534. Strong in universities, schools and training colleges. Some members of the Society supported Galileo but afterwards turned against him.

Jordan German mathematician who worked in the early stages of quantum mechanics.

Joule, James Prescott (1818–89) English physicist. Worked on the early stages of thermodynamics. The unit of energy is called after him.

Jupiter Largest planet in the solar system. Has a number of moons discovered by Galileo.

Just War According to moral philosophy a war in which certain criteria are met, making the war justifiable.

Jutland The largest naval battle of the First World War. The result was indecisive. British losses were greater, but the German fleet retired to harbour for the rest of the war.

Kaufmann, Walter German physicist. Discovered the electron at about the same time as J J Thomson. Because of his positivist outlook did not recognize it as a particle.

Kelvin, Lord (William Thomson) (1824–1907) British physicist born in Belfast but moved to Scotland as an infant. Made important advances in heat and thermodynamics, and also contributions to transatlantic cables and other practical matters.

Kepler, Heinrich Johannes Kepler's father. Mercenary soldier.

Kepler, Johannes (1571–1630) One of the greatest of all astronomers. Discovered the three fundamental laws of planetary motion.

Kepler, Katherine Johannes Kepler's mother, tried for witchcraft but released.

Kinetic theory Model of molecular motion, particularly in gases. Very successful in its early stages, but ran into difficulties which had to be resolved by quantum theory.

Koestler, Arthur (1905–83) Hungarian author and playwright. Wrote many books and newspaper articles, including a study of Copernicus, Galileo and Kepler.

Koppernigk, Niklaus (1473–1543) Polish version of Copernicus' name.

Lambda meson Apparent light meson deduced from cosmic-ray experiments in the 1940s, having a mass of about 75 electron masses. The effect was found to be due to electrons scattered from the floor of the laboratory.

Landau, Lev (1908–68) Russian physicist. Wrote many papers on low-temperature physics, and one on meson production. Nobel Prizewinner in 1962.

Langsdorf, Fritz German naval captain of the battleship *Graf Spee*, engaged with a British and New Zealand squadron in December 1939. Treated his captives with great chivalry. Committed suicide in Montevideo when ordered by Hitler to scuttle the ship.

Lansdale, Colonel Officer in the Army Security Service attached to the Manhattan Project. Gave evidence against Oppenheimer.

Lavoisier, Antoine (1743–94) French chemist. Discovered oxygen, and made other discoveries, some relevant to atomism. Guillotined during the French Revolution.

Lawrence, Ernest Inventor of the cyclotron and a senior physicist in the Manhattan Project. Nobel Prize in 1939.

Leavitt, Henrietta (1886–1921) American woman astronomer. Worked on the period–brightness relation of Cepheid variable stars, which was important in determining stellar distances.

Lemaitre, Georges (1894–1966) Belgian cleric and cosmologist. His work on the primeval atom focused attention on the big-bang model.

Leucippus (c500 BC) The first known proponent of atomic theory in the western world. His work was developed by Democritus.

Lewis, Thomas (c1720) Author of a pamphlet on Hypatia.

Liberator Consolidated B24. Long range heavy bomber. Used with great success by both USAF and RAF for anti-submarine warfare in 1943–5.

Liddell Hart, B (1895–1970) Distinguished British strategist. Wrote many books on warfare.

Lindemann, F A (1886–1957) Later Lord Cherwell. English physicist. Professor at Oxford. Member of scientific committees. A minister in two of Churchill's governments.

Linz Town in Austria. Kepler spent some years there as Imperial Mathematician.

Los Alamos Remote site in New Mexico, chosen by Groves and Oppenheimer as weapons laboratory for the Manhattan Project. There is still a research centre there.

Loschmidt, J (1821–95) Austrian physicist. The first person to determine the size of an atom.

Lucky Dragon Japanese fishing boat. Showered with radioactive ash following an American H-bomb explosion. One of the crew died and others were injured.

Lucretius (c99–55 BC) Roman philosopher. Improved the atomic theories of Democritus and Epicurus.

Luther, Martin (1483–1546) German theologian. Scandalized by the sale of indulgences for the re-building of St Peter's; and convinced that salvation was by faith alone, he drew up a list of 95 theses, which initiated the Reformation in Germany.

Lysenko, T D (1898–1976) Russian agricultural scientist. Believed in Lamarckian rather than Mendelian theory. Very influential under Stalin, he was dismissed by Kruschev.

Mach, Ernst (1838–1916) Austrian physicist and positivist philosopher. Did important work on the flow of fluids, but his insistence on direct observation as the only criterion of scientific meaning damaged physics, particularly kinetic theory and statistical mechanics.

Mach's principle The principle that the inertia of a body is due to the rest of the matter in the universe. Still controversial.

Maestlin, Michael Kepler's teacher at Tübingen, introducing him to Copernicanism.

Magellanic clouds Clusters of stars close to but not in our Galaxy.

Magnetic monopole Particle postulated by Dirac, carrying an isolated magnetic charge. Features in grand unification theory.

Manhattan Project Code name for the atomic bomb project in the US in the Second World War.

Marburg Distinguished German university, the first protestant one to be founded.

Maric, Mileva Einstein's first wife, a Serbian mathematician.

Marie Celeste The name in religion of Galileo's elder daughter Virginia.

Mars A terrestrial planet like the Earth. Its orbit was used by Kepler to derive his first law. Later in the seventeenth century it was used to set a scale on the solar system by an aberration method.

Maxwell, James Clerk (1831–79) One of the greatest physicists of the nineteenth century, making contributions in electromagnetism, kinetic theory, and other areas.

McCarthy, Joseph (1909–57) American Senator. Turned Congressional committee hearings into show trials directed at communists, liberals, the Army, State Department and finally President Eisenhower himself. After this he fell from power and soon after died.

Melancthon, Philip (1497–1560) German theologian. Composed the Augsburg confession of faith.

Mercury The innermost planet. Anomalies in its orbit were explained by general relativity.

Meson A particle with mass intermediate between the electron and the proton.

Messier, Charles (1730–1817) French astronomer. Specialized in comets and constructed a table of nebulae to aid this.

Michelson, A A (1852–1931) American optical physicist. With collaborators, made important measurements which influenced the rise of special relativity. Nobel Prizewinner in 1907.

Microsecond Millionth of a second. One thousand nanoseconds.

Microwave background Discovered by Penzias and Wilson in 1965. Powerful evidence in favour of the big bang.

Microwave radar High-powered and accurate system which can use small airborne antennae. Important in anti-submarine war.

Midshipman Junior officer in the Royal Navy. Now terminated.

Millikan, R A (1868–1953) Determined the charge on the electron and studied cosmic rays. Nobel Prizewinner in 1923.

Milton, John (1608–74) English poet. Visited Galileo in Florence after the latter's condemnation.

Mnemonics The art of improving the memory. Practised by Giordano Bruno.

Mocenigo, Giovanni Venetian merchant who denounced Bruno to the Inquisition.

Morgan, Thomas President of the Sperry Corporation and member of the AEC Board which investigated Oppenheimer.

Morley, E W American physicist. Collaborated with Michelson in their experiment to detect the ether. Their negative result led to the special theory of relativity.

Morrison, Philip American physicist, who amongst other contributions invented x-ray and gamma-ray astronomy.

Multiplicatione Specierum Roger Bacon's treatise on the propagation of light.

Muons Charged but otherwise inert particles with mass between the electron and the proton. No neutral version.

Multiple production Postulated mode of particle production. Occurs in meson production but only rarely in electron interactions.

Museum The great University of Alexandria. Founded in 323 BC, lasted until 640 AD.

Mustang American long-range fighter aircraft. Important in the closing stages of the campaign against Germany in the Second World War.

Mysterium Cosmographicum Kepler's first book, in which he puts forward his theory of the Eucliean solids and the planetary orbits.

Nagasaki Japanese city. Target for the second atomic bomb dropped on Japan in 1945.

Nanking Chinese city. Bombed by the Japanese in 1937.

Napier, John (1550–1617) Scottish mathematician and inventor of logarithms. The invention caused Kepler to rewrite his *Rudolphine Tables*.

Napoleon (1769–1821) Corsican soldier and Emperor of France.

Nature Scientific journal. Specializes in short communications in all the sciences, and in comment.

Necromancy The attempt to obtain knowledge of the future by conjuring up spirits of the dead.

Neddermeyer, Seth American physicist. Did important work on cosmic rays before the Second World War. Joined the Manhattan Project early.

Neoplatonism School of philosophy in the third to the sixth century AD. It influenced later philosophers. General title for mystical philosophy.

Nernst, Walter (1864–1941) German physicist and chemist. Made important advances in thermodynamics. Nobel Prize for chemistry in 1920.

Nestorians Followers of Nestorius (450 AD) who held that Jesus had two natures, one human, one divine. Expelled by the eastern church. His followers took knowledge of Greek civilization to the Arabs.

Niccolini, Caterina (1598–1676) Wife of the Florentine ambassador in

Rome. Cousin of Niccolo Riccardi, who was responsible for issuing Galileo's imprimatur.

Niccolini, Francesco (1584–1650) Florentine Ambassador in Rome and friend of Galileo.

Nichols, K D As colonel, asistant to General Groves in the Manhattan Project. Later general and manager of the AEC.

Nitze, Paul American economist. Collaborated with J K Galbraith in a study of bombing in Germany in 1945.

Nobel Prizes Annual prizes first awarded in 1901, endowed by Alfred Nobel. They are awarded in physics, chemistry, physiology and medicine, economics and peace.

Novelties Roger Bacon was condemned for introducing 'novelties'. These were probably scientific but may also have been philosophical and theological.

Nunn May, Alan British physicist convicted in 1946 of spying for the Soviets. He had worked on bomb projects in Montreal and Chicago.

Oboe A radar guidance system for bomber aircraft used towards the end of the Second World War.

Octahedron Eight-sided regular solid, used by Kepler in his first cosmological model.

Olbers, H W (1758–1840) Formulated an earlier paradox: why is the sky dark at night? Conclusion was that the universe is finite or expanding.

Operational research The use of mathematical models to predict the outcome of a set of initial conditions, particularly in warfare.

Oppenheimer, Frank Younger brother of Robert. Worked in a relatively junior capacity in the Manhattan project.

Oppenheimer, J Robert (1904–67) American physicist. Director of the Los Alamos laboratory, and member of the scientific panel of the Interim Committee.

Oppenheimer, Katherine 'Kitty' Wife of Robert. Had been a Communist Party member.

Opus Majus Roger Bacon's greatest work, encompassing optics, language, experimental science and observational optics (particularly the rainbow) and moral philosophy.

Opus Minus Supplement to the *Opus Majus*.

Opus Tertium Introduction to the *Opus Majus*.

Orestes Roman prefect in Alexandria c400 AD. Friend of Hypatia.

Osborne Former junior naval college, preparatory to Dartmouth.

Osiander German Lutheran cleric. Wrote a spurious preface to Copernicus' *De Orbis*.

Ostwald, F W (1853–1932) German chemist and positivist, allied to Mach. Nobel Prizewinner in Chemistry 1909.

Padua City under the control of Venice where Galileo spent the years 1592–1610.

Pair production The generation of a positive and a negative electron by a high-energy gamma ray.

Paleomagnetism The study of the direction and strength of magnetism in rocks. Important in the study of continental drift.

Palatinate State of the old German empire, west of the Rhine. Partitioned in 1815.

Pantheism The belief that God is resident in the universe and composed of the universe.

Parabolans Para-military militia in fifth-century Alexandria.

Paradigm All the factors which influence the work of the scientist. Term introduced by T Kuhn.

Parallax Apparent motion of a star relative to nearby ones, as the Earth goes through its orbit. A similar method can be used for planetary measurements.

Pasadena Town in California and home of the California Institute of Technology.

Pash, Colonel Boris American security officer during the Second World War.

Pathfinder Method of organizing bombing raids in the Second World War. High-speed, usually Mosquito aircraft, with a low radar profile, guided the main wave of bombers to the target by dropping flares. Pathfinder aircraft would remain over the target for some hours.

Pauli, W (1900–58) Austrian physicist. Originator of the exclusion principle. Nobel Prizewinner in 1945.

Paul III, Pope (1468–1549) Revived the Inquisition, excommunicated Henry VIII and called the Council of Trent.

Paul V, Pope (1552–1621) Notoriously gloomy Pope. Broke with Venice. Gave audience to Galileo.

Perrin, Jean (1870–1942) French physicist. Studied Brownian motion of small particles in liquids, demonstrating the existence of atoms. Nobel Prizewinner in 1926.

Pion Nuclear active meson. Mass 274 electron masses for the charged particles, 255 for the neutral.

Planets Relatively small objects rotating around a star. They may have satellites themselves, as with Jupiter and the Earth.

Plato (428–348 BC) Perhaps the most influential of all philosophers, but with a smaller influence on science. His emphasis on mathematics was important.

Plural production Successive collisions with separate targets, one secondary particle being produced at a time. See **Multiple production**.

Plutonium 239 Man-made element derived from the bombardment of U238 with neutrons. Can be made into an atomic bomb, or used as fuel for a nuclear reactor.

Podolsky, B With Einstein and Rosen wrote a paper in 1935, challenging the accepted interpretation of quantum mechanics.

Poincaré, J H (1854–1912) French mathematician. His work was important in the development of relativity.

Positivism Philosophical system that confines knowledge to what is observed. Gave rise to a bitter controversy over statistical mechanics in the nineteenth century.

Positron Positive electron predicted by Dirac and discovered by Anderson.

Powell, C F (1903–69) Leader of a very successful group at Bristol University in the 1940s and 1950s. Discovered several mesonic particles. Nobel Prizewinner in 1950.

Prague Imperial capital in the seventeenth century.

Primordial black holes If the universe is sufficiently chaotic in its early stages these may be produced with a wide spectrum of masses.

Primum mobile First and outermost sphere in the Ptolemaic model.

Proton Basic constituent of the atomic nucleus. Mass 1836 times the electron mass.

Providentissimus Deus Encyclical issued by Pope Leo XIII in 1893. Important because it vindicates Galileo's views on the interpretation of Scripture.

PSB Personnel Security Board. The committee set up in 1954 to review Oppenheimer's security clearance.

Ptolemy, Claudius (c90–168 AD) The most influential astronomer of antiquity. His geocentric model of the solar system held sway until the sixteenth century.

Pugwash Pacifist movement set up by Einstein and Bertrand Russell after the Second World War.

Pulley, The Title of a theological poem by George Herbert often quoted by J R Oppenheimer.

Pythagoras (sixth-century BC) Greek philosopher and mathematician. Influenced Plato and others.

QSO Quasi stellar object. Powerful but very small celestial objects, some visible at distances close to the edge of the universe.

Quantum mechanics The new theory propounded in 1925–7, by Heisenberg, Dirac and others.

Quantum theory Term now generally used for the 'old' quantum theory, i.e. that current from 1900 to 1924.

Rabi, I I (1898–1988) American physicist. Nobel Prizewinner in 1944.

Radar, centimetric Short-wave radar made possible by the cavity magnetron. Important in the Battle of the Atlantic.

Radiation length In an electron–photon cascade, the average length of each generation. Varies with the medium.

Radio astronomy Founded by Karl Jansky, the study of the universe at radio frequencies.

RAE Royal Aircraft Establishment, Farnborough.

Rapporteur papers Summary papers which are read at conferences. Provide opportunities for dialogue and possibly conflict.

Rayleigh, John William Strutt, 3rd Baron (1842–1919) English physicist. Worked in a variety of areas of classical physics. Nobel Prizewinner in 1904.

Reber, Grote American radioastronomer. Followed Jansky in pioneering work on the radio sky.

Red-shift The change in wavelength of spectral lines as a galaxy moves away from us.

RFC Royal Flying Corps. Predecessor of the RAF.

Reformation A process lasting a hundred years, wherein large parts of Europe broke away from the Roman Church. After several abortive attempts over the previous two hundred years, in the early sixteenth century Martin Luther successfully challenged the Pope on indulgences and other issues.

Regensburg City in Bavaria. Kepler died there in 1630.

Riccardi, Niccolo (1585–1639) Master of the Sacred Palace under Urban VIII. Issued one of the imprimaturs for Galileo's *Chief World Dialogues.*

Robb, Roger Counsel for the AEC in the matter of J R Oppenheimer.

Rochester City in upper New York State. In physics its university is noted for optics and nuclear physics.

Roosevelt, Franklin Delano (1882–1945) American President. Coming into office during the depression, he introduced a series of radical measures to combat poverty and create employment. He brought the USA into the Second World War but died before its end. He authorized work on the atomic bomb.

Rosen, Nathan Collaborated with Einstein and Podolsky on the paper of 1935, which questioned the accepted interpretation of quantum mechanics.

Rosenbergs, Ethel and Julius Americans, executed for espionage in 1951.

Rossi, Bruno Carried out pioneering work on cosmic rays in the early 1930s.

Royal Society of London Premier scientific academy in Britain. Chartered in 1660.

Rubens, Otto German infrared spectroscopist. His inability to detect N-rays led to the collapse of the subject.

Rudolphine Tables Astronomical tables commissioned by the Emperor Rudolph and completed after many years by Kepler. Published in 1627.

Rutherford, Ernest (1871–1937) New Zealand physicist. Pioneer of nuclear physics; discovered the nuclear atom and the disintegration of the nucleus. Nobel Prize for Chemistry, 1908.

Sagredo, Salviati and Simplicio The interlocutors of Galileo's *Two World Dialogue*. Sagredo and Salviati were based on real characters. It was suggested to Urban VIII that Simplicio was based on him. This prejudiced him against Galileo.

Sarpi, Paolo (1522–1623) Venetian patriot and historian. The guiding spirit behind Venetian autonomy. Carried out research in anatomy.

Saturn One of the major planets. Featured in Kepler's work on the Euclidean solids.

Scheele, C W (1742–86) Swedish chemist. Discovered many acids. Promoted atomic theory.

Scheiner C (1573–1650) German Jesuit astronomer. Possibly discovered sunspots. In conflict with Galileo.

Schelling, T C Author of *The Strategy of Conflict* (1960).

Scholasticism Synthesis of the thought and theology of the schoolmen, from St Augustine and Anselm, but particularly from Albert the Great and Aquinas in the thirteenth century.

Schrödinger, Erwin (1887–1961) Austrian physicist. His career was interrupted by the rise of Nazism, but he settled in the Dublin Institute for Advanced Studies from 1940 to 1956. Best known for his wave equation, he also worked in statistics and unified field theory.

Science American scientific journal with letters and comment.

Scientific Panel A subcommittee of the Interim Committee designed to give scientists some input into decisions about the use of the atomic bomb.

Scripture During the mediaeval period thought to be literally true, though St Augustine and others wrote on its figurative interpretation. Became a major issue with Galileo.

Seaborg, Glenn American chemist. Discovered plutonium, americium, and curium. Later berkelium and californium. Nobel Prizewinner in 1951. Chairman of the AEC, 1961–71.

Seghizi de Lauda Commissary of the Inquisition at the time of Galileo's visit to Rome in 1615–16.

Shapley, Harlow (1885–1972) American astronomer. Showed that the solar system was not at the centre of our Galaxy.

Sidereus Nuncius (*The Starry Messenger*) Pamphlet by Galileo describing his discoveries with a telescope in 1610.

Sidney, Sir Philip, (1554–86) English poet and soldier. Much loved for his courage, generosity, and chivalry. Died of wounds received at the Battle of Zutphen. Patron of Giordano Bruno.

Singularity A point of infinite density and zero size. Predicted at the origin of the universe by general relativity, but a quantum theory of gravity is needed to confirm or deny this.

Sirius Bright star, used by Huygens in his measurement of the distance from the Earth to a star.

Snow, C P British administrator. Originally trained as a physical chemist, he entered the Civil Service, specializing in the selection of scientific personnel. Wrote many novels and some essays, notably on the Blackett–Cherwell controversy.

Solar distance First measured by Aristarchus. His value was too small but much more realistic than previous estimates. Measured more accurately in the seventeenth century.

Spanish loyalists Supporters of the Republic during the Spanish Civil War, 1936–9.

Specific heats C_p, C_v The specific heats at constant pressure and volume. Their ratio is important in the kinetic theory of gases. Anomalies were finally explained by quantum theory.

Sphere One of the components of Kepler's cosmological model.

Spiral nebulae Our own and many other galaxies. Usually they have a multiple spiral structure and contain about a hundred billion stars.

Stefani, G (1577–1633) Florentine Professor Consultant to the Inquisition for Galileo's *Chief World Dialogues*.

Stimson, Henry Secretary of War in Truman's Cabinet. Advised the President on the use of the atomic bomb.

Stoicism Philosophical system founded about 300 BC, Anti-atomist.

Stokes, G (1819–1903) Irish physicist and mathematician. Most important work was on the viscosity of solids and liquids.

Stokes, Richard (MP) British politician who during the Second World War, questioned the wisdom and morality of Allied bombing policy.

Strauss, Admiral Chairman of the AEC in 1953.

Summa Theologia Best known and most influential work of St Thomas Aquinas (1225–74). Originally intended as a student summary, it was the dominant work in the Roman church until this century.

Sunspots Originally discovered in China, their priority in the west was disputed between Scheiner and Galileo. Important because they demonstrated change in the celestial sphere.

Super Colloquial name for the hydrogen bomb.

Supernova Stars which having reached a terminal state, explode, releasing heavy elements into interstellar space. Can be used for distance measurements.

Supersymmetric particles Postulated particles which match each known particle, e.g. photinos, gravitinos.

Suzuki, Admiral Japanese premier in 1945 and member of the peace faction.

Swabia South German state. Kepler lived in Ulm and Einstein was born there.

Sweden In the seventeenth century one of the most aggressive states. Played an important part in the early part of the Thirty Years War.

Swift, Jonathan Possessed a copy of Sarpi's *History of the Council of Trent*; indicating the importance attached to it by Protestant scholars even a hundred years later.

Synchrotron radiation Radiation emitted when electrons gyrate in a magnetic field. Most of the galactic radio background is due to this process.

Synesius Pupil of Hypatia, who afterward became a bishop. His letters are an important source of information about her.

Szilard, Leo Hungarian engineer, who worked with Einstein, Fermi and other physicists. Credited with the invention of the chain reaction, though not through neutrons.

Target Committee A sub-committee of the Interim Committee charged with finding a mixed military and civilian site suitable for an atomic bomb target.

Tatlock, Jean Mistress of J R Oppenheimer. Committed suicide in 1944.

Teller, Edward (1908–) Hungarian-American physicist. Strong proponent of the hydrogen bomb.

Tengnagel von Camp, Franz Son-in-law of Tycho Brahe.

Terrestrial matter In Aristotelian cosmology, base, earthy material.

Tesla, Nikola (1856–1943) Yugoslav physicist and engineer. Worked with Edison but refused to share the 1912 Nobel Prize with him.

Tetrahedron Four-sided regular solid used by Kepler in his cosmology.

Thales (c625–545 BC) Early Greek philosopher, the first to think scientifically about the universe.

Theon c370 AD Greek philosopher and astronomer, father of Hypatia. Wrote commentaries, some of which have survived, about Ptolemy and other authors.

Thermonuclear reaction Basically the combination of four nucleons to make helium. In practice the combination of deuterium and tritium is the most feasible.

Third law of planetary motion The square of a planet's year is proportional to the cube of the mean distance from the Sun. Discovered by Kepler in 1618, but not understood for several decades.

Thirty Years War (1618–48) Commenced as a religious conflict, spread to engulf most of central Europe. Caused more deaths than any conflict before the First World War.

Thomson, J J (1856–1940) English physicist. Discoverer of the electron.

Tides Studied by Galileo, who saw them, mistakenly, as strong evidence for the Copernican model.

Tokyo Scene of the greatest air raid in history.

Toland, John Published in 1720 a long historical essay on Hypatia, attacking Cyril, the Patriarch of Alexandria.

Towneley, Richard Seventeenth-century Lancashire gentleman, who assisted in the foundation of Boyle's Law for gases.

Transcript Title of the verbatim record of the proceedings in the case of J R Oppenheimer.

Trenchard, Sir Hugh (later Viscount) Founder of the RAF.

Tritium Isotope of hydrogen with one proton and two neutrons. Feebly radioactive.

Trypanosoma cruzi Organism which if injected into the bloodstream, causes debility and sometimes death. Darwin's illness has been attributed to it.

Tübingen German university town. Michael Maestlin, a Professor there, taught Kepler the Copernican model.

Tuscany Independent Duchy, with Florence as its capital. Galileo lived there for thirty years.

Two New Sciences Galileo's last and perhaps greatest work, published in Holland in 1638. It lays the foundations of mechanics.

Tycho See **Brahe**.

U235 The isotope of uranium with total mass 235 units. Occurs as less than 1% of natural uranium.

U238 The isotope of uranium with total mass 238 units. Over 99% of natural uranium.

Ulm City in Swabia, South Germany. Birthplace of Einstein.

Urban VIII, Pope Formerly Maffeo Barberini. Friend and patron of Galileo but afterwards turned against him.

Vallarta Mexican physicist. Collaborating with Lemaitre he showed that the primary cosmic radiation must be charged.

van der Waals Dutch physicist. Worked on departures from the ideal gas laws.

Vavilov Russian geneticist and opponent of Lysenko. Disappeared in one of Stalin's purges.

Venice Independent republic in the seventeenth century. Also controlled Padua.

Venus' gibbous phase The phase between half and full circle illumination. Can only occur if Venus rotates about the Sun.

Viscosity of gases Predicted by Maxwell to be independent of pressure. Strong evidence for the kinetic theory of gases.

Wallenstein (1583–1634) Austrian General in the Thirty Years War. Initially successful, he was assassinated in 1634.

Wehrmacht The German Army.

White holes The inverse of black holes, emitting matter and radiation instead of absorbing it.

Wilberforce, Bishop Samuel (1805–73) Controversialist, figuring large in the debate with T H Huxley over evolution.

William of Auvergne (died 1249) Roger Bacon heard him lecture. This puts some limits on the chronology of Bacon's stay in Paris.

Wilson, C T R (1869–1959) Scottish physicist. Invented the cloud chamber, with which many important discoveries were made. Nobel Prizewinner in 1927.

Wittenberg German university town where Martin Luther was a Professor.

Wood, R W (1868–1955) American physicist specializing in optics. His investigation of N-rays led to the collapse of the subject.

Yukawa, Hideki (1907–81) Japanese physicist. Postulated an exchange theory of nuclear forces. Its particle would be intermediate in mass between the electron and proton, estimated at about 200 electron masses. Though opposed by Bohr it was eventually accepted.

Zeppelin German airships which bombed London during the First World War.

Zuniga, Diego de Wrote a commentary on the Book of Job, Copernican in tone. Condemned by the Inquisition in 1616.

BIBLIOGRAPHICAL DISCUSSION

2a. Hypatia

None of Hypatia's works survive, though it is generally believed that her edition of Ptolemy's *Almagest* is largely that which we have today. Diophantus' work *Arithmetika* probably owes its existence to the quality of her editing. Her commentary on Apollonius' important work on the conic sections has been reconstituted in modern times [Perl T 1978 *Math. Equals: Biography of Women Mathematicians* (Menlo Park, CA: Stanford)]. This was of course an important source for Kepler.

Her father Theon has fared somewhat better, though he was not as distinguished as his daughter. His commentaries were on Euclid and Ptolemy. Original work was largely confined to allegorical poetry, which still survives. He wrote a book on the astrolabe which has not survived but has been largely reconstituted from Arab sources. Outside astronomy he was interested in biology and occult poetry. He was an active astrologer.

Sources

The most important early mention of Hypatia comes from the ecclesiastical history of Socrates Scholasticus (379–450) [Schaff P and Wace H (ed) 1952 *The Ecclesiastical History of Socrates Scholasticus: Fathers of the Christian Church* (Michigan: Michigan University Press) vol ii, pp viii–xvi]. There is extensive writing on this source both in early and modern times. Her pupil, bishop Synesius, left six letters which have survived. Most historians have used the *Suda*, a tenth-century Byzantine encyclopaedia. Hesychius of Milètus, a sixth-century historian, is the main source of the Hypatia entry in the Suda. A number of sources have been collected by Dzielska [Dzielska M 1995 *Hypatia of Alexandria* (Cambridge, MA: Harvard University Press)].

2b. Roger Bacon

His major texts have survived and have been completely edited and translated in this century. The *Opus Majus*, his greatest work, has received particular attention. There is a two-volume translation by Burke [Burke R B 1962 *The Opus Majus of Roger Bacon* (New York: Russell and Russell)]. The important work, *Scriptum Principale,* however, is fragmentary.

Bacon refers in glowing terms to the work of Peter of Maricourt, who was expert in magnetism and other physical phenomena. There are fragmentary references to him elsewhere, and three centuries later Gilbert discusses his work, but otherwise nothing is known.

Bacon wrote other works such as *Multiplicatione Specierum* (*Multiplication of Species*) and the *Compendium.* With the *Opus Tertium* and the *Opus Secundum* these tend to be introductions or supplements to the *Opus Majus.*

Much of his work is obscure and he shows an extraordinary credulity side by side with detailed scientific work. Particularly important for his science is Lindberg's book [Lindberg D C 1983 *Roger Bacon's Philosophy of Nature* (Oxford: Clarendon)]. Bacon's work had large popular interest and influenced later writers. The fifteenth-century writer Pierre d'Ailly, gives a long discussion of the geographical relations of Spain and India, which is lifted from the *Opus Majus.* John Dee in 1582 praised Bacon's work in a memorial addressed to Queen Elizabeth.

2c. Giordano Bruno

Bruno's writing is allusive and allegorical, but much of it has survived, since he lived in the era of the printed book. It consists of very many controversial pamphlets, three important works in Latin: *De Immenso, De Monade,* and *De Minimo.* In Italian he wrote *Lo Spaccio* (*The Expulsion of the Triumphant Beast*), *La Cena de le Ceneri* (*Ash Wednesday Supper*), and five other works. Some of the pamphlets have disappeared, but the major works are still extant.

Cena de le Ceneri, and *De l'Infinito et Mondi,* are of particular interest to the scientist; the first is a statement of the Copernican model and its consequences; the second a study of infinite universes. The latter warrants Bruno being credited with the Copernican Principle.

All Bruno's works are written in dialogue form and most are satirical in character. It is an immense output and very varied. Though totally unlike modern scientific writing, it states important scientific principles for the first time since Aristarchus. This seems to me to warrant including Bruno with Galileo and Kepler as a genuine scientist, though not of course of the same order.

3. Galileo

As one of the major figures of the seventeenth century, Galileo commands an enormous literature. His works are available in book form, mostly in their original editions. These have been collected and edited by Antonio Favaro, in his monumental National Edition, comprising twenty volumes, and taking thirty years, from 1880 to 1910, to complete. This contains Galileo's works, letters and documents relating to him.

The first English translation of the *Two Worlds Dialogue* was made by Thomas Salusbury in the seventeenth century. It has been revised by de Santillana [de Santillana 1953 (Chicago, IL: University of Chicago Press)]. A modern translation has been made by Stillman Drake [Drake S 1953 *The Two Chief World Systems* (Berkeley, CA: University of California) with a foreword by Albert Einstein].

Other principal works by Galileo were:

The Starry Messenger, 1610
Letters on Sunspots, 1613
Letter to the Grand Duchess Christina, 1615
The Assayer, 1623

All the above have been translated into English by Stillman Drake [Drake S 1957 *Discoveries and Opinions of Galileo* (New York); Galileo G 1638 *Dialogues on Two New Sciences* (Amsterdam: Elzevir) Engl. Transl. Crew H and de Salvio A 1914 (New York: Dover) and 1974 Drake S (Madison: Wisconsin University Press)].

Some Books on Galileo in English

Wallace W A 1977 *Galileo's Early Notebooks: The Physical Questions* (Notre Dame: Notre Dame University Press)

Wallace W A 1984 *Galileo and His Sources* (Princeton, NJ: Princeton University Press)

de Santillana G 1958 *The Crime of Galileo* (Heinemann)

Brodrick J 1964 *Galileo* (London: Chapman)

Koestler A 1959 *The Sleepwalkers* (New York: Macmillan)

von Gebler K 1879 *Galileo Galilei and the Roman Curia* (London)

Fahie J J 1903 *Galileo, His Life and Work* (London)

Kuhn T S 1956 *The Copernican Revolution* (Cambridge, MA: Harvard University Press)

Finocchiaro M (ed) 1989 *The Galileo Affair* (Berkeley, CA: University of California) (documentary history containing the relevant letters to or from Galileo)

4. Kepler

Kepler's astronomical output is smaller than that of some of his contemporaries, but his books are major works. In 1597 he published *Mysterium Cosmographicum* (the mystery of the universe), printed at Tübingen. This is today extremely rare to find, and accessible only in modern editions. Its planetary illustrations are drawn to scale, which makes some of them enormous.

In 1609 came the *Astronomia Nova* (*The New Astronomy*) printed in Heidelberg. Again, copies of the original printing are very rare, although the book contains the first two laws of planetary motion.

In 1618–21 the *Epitome of Copernican Astronomy* was published. It was banned by the Inquisition. It was followed (1619) by the *Harmonice Mundi* (*Harmonies of the World*), which contains the third law and is Kepler's crowning achievement.

Considered more important at the time was the *Rudolphine Tables*, printed in Ulm and published in Frankfurt.

Other works include *On the More Certain Fundamentals of Astrology* [Engl. Transl. Rossi M 1976 (Madison: Wisconsin University Press)]. In technical and non-scientific work he was quite prolific, publishing theological works and practical treatises on the measurement of the volume of wine-barrels and similar topics.

The complete works in Latin were edited by C Frinck, and a German edition was produced by Max Caspar (Munich) in 1937. This was translated into English in 1959.

Books and Chapters in Books

Boas M 1962 *The Scientific Renaissance* (London: Collins)

Joannes Kepler. Engl. Transl. Max Caspar 1959

Beer A and Beer P 1975 *Kepler: Four Hundred Years (Vistas in Astronomy 18)* (Oxford: Pergamon)

Kozhamthadam J 1994 *The Discovery of Kepler's Laws* (Notre Dame: Notre Dame University Press)

Kuhn T 1959 *The Copernican Revolution* (New York: Vintage Books)

Knight D C 1965 *Johannes Kepler* (London: Chatto and Windus)

Articles

Wilson C A 1968 Kepler's derivation of the elliptical path *Isis* **59** 5–25

Aiton E J 1969 Kepler's second law of planetary motion *Isis* **60** 75–90

Gingerich O 1972 Kepler and the new astronomy *Quart. J. R. Astron. Soc.* **13** 346–73

5. Boltzmann

Biographies of the principal characters are given in the text. The crucial argument over degrees of freedom is described in a contemporary lecture by Kelvin [Kelvin 1904 *Baltimore Lectures* (London: Clay)] an extract from which is in the appendix to Chapter 5. It is, however, difficult for a modern reader to follow because of the unfamiliar style of nineteenth-century scientific writing. All the significant papers and books in this controversy are still extant.

6. N-Rays

The best description of the development of N-rays, in the English language, are those given in volumes of *Nature* from 1903–5. Some of these are given in the text of Chapter 6.

7. Einstein

His developing views on quantum theory are given in many books, but apart from his papers, he wrote rather little in books himself. The bibliography given by P A Schilpp amounts to 62 pages, and this is the most comprehensive source for the controversy:

Schilpp P A (ed) 1949 *Albert Einstein: Philosopher Scientist* (Evanston, IL: Library of Living Philosophers)

See also:

1961 *Relativity* (New York: Crown) (first published 1916)

Prziban K (ed) 1967 *Letters on Wave Mechanics* (New York: Philosophical Library)

Einstein A 1950 *The Meaning of Relativity* (Princeton, NJ: Princeton University Press)

Einstein A 1950 *Out of My Later Years* (New York: Philosophical Library)

Einstein A *et al* 1952 *The Principle of Relativity* (New York: Dover)

Einstein A and Infeld L 1938 *The Evolution of Physics* (Cambridge: Cambridge University Press)

8. Oppenheimer

He wrote rather few books, mainly in collections or the result of a lecture series. He published 48 papers in *Physical Review* or *Physical Review Letters*.

Books

Oppenheimer J R 1954 *Science and the Common Understanding (BBC Reith Lectures 1953)* (Oxford: BBC)

Oppenheimer J R 1955 *The Open Mind* (New York: Simon and Schuster)

Oppenheimer J R 1956 *The Constitution of Matter* (Oregon State University)

Books on Oppenheimer or related subjects

1954 *The Trial: In the Matter of J Robert Oppenheimer. Transcript of Hearings before the Personnel Security Board, US Atomic Energy Commission* (Washington, DC)

Rhodes R 1986 *The Making of the Atomic Bomb* (London: Penguin)

Groves L 1963 *Now it Can be Told* (London: Deutsch)

Reid R W 1971 *Tongues of Conscience* (London: Panther)

Dyson F 1993 *From Eros to Gaia* (London: Penguin)

Rouze M 1964 *Oppenheimer* (London: Souvenir)

9. Blackett, Cherwell and Tizard

This chapter has diverse topics and people. Brief biographies of the principal characters are given at the beginning of Chapter 9. These are drawn from C P Snow's article, Ronald Clark's biography of Tizard and recollections by friends.

Books on Ethics in War

Gill R 1985 *A Textbook on Christian Ethics* (Edinburgh: Clark)

Bonhoeffer D 1966 *Ethics* (London: Fontana)

Anscombe E 1948 *Mr Truman's Honorary Degree* Pamphlet, printed privately

Walzer M *Just and Unjust Wars* (New York: Basic Books)

Strategy and Operational Research

Churchill W S 1951 *History of the Second World War* vols 3, 4, 5 (London: Cassell)

Feis H 1961 *Japan Subdued* (Princeton, NJ: Princeton University Press)

Liddell Hart B 1970 *The History of the Second World War* (London: Cassell)

Irving D 1963 *The Destruction of Dresden* (London: Kimber)

Snow C P 1971 *Science and Government. In Public Affairs* (London: Macmillan)

Frankland N 1965 *The Bombing Offensive against Germany* (London: Faber and Faber)

Clark R W 1965 *Tizard* (London: Methuen)

Blackett, P M S see Snow *op. cit.* p 179, Clark *op. cit.* p 309

Dyson F 1992 *From Eros to Gaia* (London: Penguin)

10. The Big Bang

Silk J 1989 *The Big Bang* (New York: Freeman)

Barrow J D and Silk J 1993 *The Left Hand of Creation* (London: Penguin)

Weinberg S 1983 *The First Three Minutes* (London: Fontana)

Contopoulos G and Kotsakis D 1987 *Cosmology* (Berlin: Springer)

Sciama D W 1971 *Modern Cosmology* (Cambridge: Cambridge University Press)

Arp H, 1987 *Quasars, Redshifts and Controversies* (Berkeley, CA: University of California)

11. Multiple versus Plural

Blackett P M S and Occhialini G P S 1933 The soft cascade *Proc. R. Soc.* **139** 699

Anderson C D *et al* 1934 *Phys. Rev.* **45** 352

Carlson J F and Oppenheimer J R 1936 *Phys. Rev.* **51** 220

Bhabha H J and Heitler W 1936 *Proc. R. Soc.* **159** 432

Heitler W 1944 *Quantum Theory of Radiation* (Oxford: Oxford University Press)

13. Conclusions

Conflict has been treated by many authors in many different contexts in this century, apart from the vast literature extending to ancient times.

Schelling T 1966 *The Strategy of Conflict* (Cambridge, MA: Harvard University Press); covers the phenomenon in a general way

Freedman L 1988 *The Evolution of Nuclear Strategy* (New York: Macmillan); is specifically directed at the Cold War

Kuhn T 1962 *The Structures of Scientific Revolutions* (Cambridge, MA: Harvard University Press); deals with conflict in science, but as a side issue.

BIBLIOGRAPHY

Anscombe E 1948 *Mr Truman's Honorary Degree* Pamphlet, printed privately

Aquinas T 1989 *Summa Theologia* (Engl. Transl.) (London: Methuen)

Arnold-Foster M 1973 *The World at War* (London: Collins)

Arp H C 1987 *Quasars Redshifts and Controversies* (Berkeley, CA: University of California)

Aspect A, Dalibaud F, Grangier P and Roger A 1981 *Phys. Rev.* **47** 460

Bacon R 1962 *Opus Majus* R D Burke (ed) (New York: Russell)

Bainton R H 1961 *Christian Attitudes to Peace and War* (London: Hodder and Stoughton)

Bernstein J 1973 *Einstein* (London: Fontana)

Bettany G T 1890 *World Religions* (London: Bracken)

Boas M 1962 *The Scientific Renaissance 1450–1630* (London: Collins)

Bonhoeffer D 1966 *Ethics* (London: Fontana)

Bridges J H 1914 *The Life and Work of Roger Bacon* (London: Williams and Norgate)

Broderick J 1964 *Galileo* (London: Chapman)

Bruno G 1964 *The Expulsion of the Triumphant Beast* (Engl. Transl. Imerti A) (New Brunswick: Rutgers University Press)

Brush S 1983 *Statistical Physics and the Atomic Theory of Matter* (Princeton, NJ: Princeton University Press)

Burbidge G 1990 *Physical Cosmology* (Paris: Editions Frontieres)

Burtt E A 1954 *The Metaphysical Foundations of Modern Science* (New York: Doubleday)

Christie-Murray D 1978 *A History of Heresy* (Oxford: Oxford University Press)

Churchill W S 1951 *History of the Second World War* vols 3–5 (London: Cassell)

Clark R W 1965 *Tizard* (London: Meuthen)

Clark R W 1973 *Einstein* (London: Hodder and Stoughton)

Compton A H 1956 *Atomic Quest* (Oxford: Oxford University Press)

Contopoulos G and Kotsakis D 1984 *Cosmology* (Berlin: Springer)

de Santillana G 1958 *The Crime of Galileo* (London: Heinemann)

Dictionary of National Biography

Dyson F J 1993 *From Eros to Gaia* (London: Penguin)

Dzielska M 1995 *Hypatia of Alexandria* (Cambridge, MA: Harvard University Press)

Einstein A, Podolsky B and Rosen N 1935 *Phys. Rev.* **47** 777

Encyclopaedia Britannica: Bacon R

Feis H 1961 *Japan Subdued* (Princeton, NJ: Princeton University Press)

Finocchario M (ed) 1989 (Berkeley, CA: University of California Press)

Flannery A 1975 *Gaudium et Spes* (Dublin: Dominican)

Flowers B and Mendoza E 1970 *Properties of Matter* (New York: Wiley)

Frankland N 1965 *The Bombing Offensive against Germany* (London: Faber and Faber)

French A P (ed) 1979 *Einstein* (London: Heinemann)

Furley D J 1967 *Two Studies in the Greek Atomists* (Princeton, NJ: Princeton University Press)

Galileo G 1914 *Two New Sciences* (New York: Dover)

Galileo G 1962 *Dialogue Concerning the Two Chief World Systems* (Engl. Transl. Drake S) (California: California University Press)

Gill R 1985 *A Textbook of Christian Ethics* (Edinburgh: Clark)

Giovannitti L and Freed E 1961 *The Decision to Drop the Bomb* (London: Methuen)

Gowing M 1964 *Britain and Atomic Energy* vol 1 (London: Macmillan)

Groueffe S 1967 *Manhattan Project* (London: Collins)

Groves L 1963 *Now it Can be Told* (London: Deutsch)

Hamilton W R 1895 *Collected Correspondence* (Dublin: Royal Irish Academy)

Hawking S W 1988 *A Brief History of Time* (London: Bantam)

Hughes P 1954 *A Popular History of the Catholic Church* (New York: Doubleday)

Jeans J 1944 *The Growth of Physical Science* (Cambridge: Cambridge University Press)

Jungk R 1958 *Brighter than a Thousand Suns* (London: Gollancz)

Kelvin Lord 1904 *Baltimore Lectures* (London: Clay)

Kepler J 1981 *Mysterium Cosmographium* (Engl. Transl. Duncan A M) (New York: Abaris Books)

Kilmister C W (ed) 1987 *Schrödinger* (Cambridge: Cambridge University Press)

Knight D C 1965 *Johannes Kepler* (London: Chatto and Windus)
Kozmathadam J 1994 *The Discovery of Kepler's Laws* (Notre Dame)
Krane K 1988 *Introductory Nuclear Physics* (New York: Wiley)
Kuhn T 1977 *The Essential Tension* (Chicago: Chicago University Press)
Liddell Hart B 1970 *History of the Second World War* (London: Cassell)
London Encyclopedia 1829
McCormmach R 1982 *Night Thoughts of a Classical Physicist* (New York: Avon)
Marshall Libby L 1979 *The Uranium People* (New York: Scribner)
Maxwell J C 1902 *Collected Correspondence* (Oxford: Oxford University Press)
Medawar P 1967 *The Art of the Soluble* (London: Methuen)
Nature 1903–4 **68–70**
Newman J H 1870 *A Grammar of Assent* (London: Burns Oates)
Newton I 1980 *Correspondence* (Cambridge: Cambridge University Press)
Ogg D 1948 *Europe in the Seventeenth Century* (London: Black)
Pais A 1982 *Subtle is the Lord. The Science and the Life of Albert Einstein* (Oxford: Oxford University Press)
Polkinghorne J C 1989 *The Quantum World* (New York: Longmans)
Redondi P 1987 *Galileo Heretic* (Princeton, NJ: Princeton University Press)
Reid R W 1971 *Tongues of Conscience* (London: Panther)
Rhodes R 1986 *The Making of the Atomic Bomb* (London: Penguin)
Rouze M 1964 *Robert Oppenheimer* (London: Souvenir)
Rowan-Robinson M 1977 *Cosmology* (Oxford: Oxford University Press)
Russell H, Dugan R and Stewart J 1945 *Astronomy* vol 1 (Boston: Ginn)
Schilpp P (ed) 1949 *Albert Einstein* (New York: Tudor)
Schelling T C 1966 *The Strategy of Conflict* (Cambridge, MA: Harvard University Press)
Sciama D W 1971 *Modern Cosmology* (Cambridge: Cambridge University Press)
Silk J 1989 *The Big Bang* (San Francisco: Freeman)
Smyth H D 1945 *Atomic Energy for Military Uses* (London: HMSO)
Snow C P 1971 *Science and Government in Public Affairs* (London: Macmillan)
Squires E 1994 *The Mystery of the Quantum World* (Bristol: Institute of Physics Publishing)
Stimson H L 1947 The decision to use the atomic bomb *Harpers Magazine*
Wallace W 1984 *Galileo and his Sources* (Princeton, NJ: Princeton University Press)
Walzer M 1977 *Just and Unjust Wars* (New York: Basic Books)

Waugh E 1987 *Men at Arms* (London: Penguin)
Weinberg S 1993 *Dreams of a Final Theory* (London: Vintage)

Index